About Island Press

Since 1984, the nonprofit organization Island Press has been stimulating, shaping, and communicating ideas that are essential for solving environmental problems worldwide. With more than 1,000 titles in print and some 30 new releases each year, we are the nation's leading publisher on environmental issues. We identify innovative thinkers and emerging trends in the environmental field. We work with world-renowned experts and authors to develop cross-disciplinary solutions to environmental challenges.

Island Press designs and executes educational campaigns, in conjunction with our authors, to communicate their critical messages in print, in person, and online using the latest technologies, innovative programs, and the media. Our goal is to reach targeted audiences—scientists, policy makers, environmental advocates, urban planners, the media, and concerned citizens—with information that can be used to create the framework for long-term ecological health and human well-being.

Island Press gratefully acknowledges major support from The Bobolink Foundation, The Curtis and Edith Munson Foundation, The Forrest C. and Frances H. Lattner Foundation, The Freedom Together Foundation, The Kresge Foundation, The Summit Charitable Foundation, Inc., and many other generous organizations and individuals.

The opinions expressed in this book are those of the author(s) and do not necessarily reflect the views of our supporters.

THE TWILIGHT FOREST

The Twilight Forest

An Elegy for Ponderosa
in a Changing West

Gary Ferguson

◐ **ISLAND**PRESS | Washington | Covelo

Library of Congress Control Number: 2025936289

All Island Press books are printed on environmentally responsible materials.

Manufactured in the United States of America
10 9 8 7 6 5 4 3 2 1

Keywords: American West, Arizona, California, Colorado, John Muir, Mogollon Rim, New Mexico, ponderosa pine, Southwest, Utah, canyon, climate change, drought, ecology, forest, forest fire, natural history, wildfire

For Mary, who, even in this troubled world,
shows me how to dance.

Contents

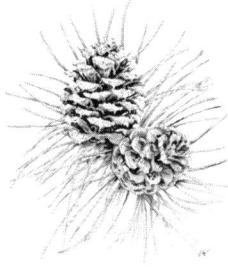

Chapter One

I 'm in the great wild open of southern Utah, standing on a slab of rock the size of a school bus, swooning over a view to the south stretching on for hundreds of square miles, clean into Arizona. A perfect garble of red rock turrets and trees and scoured canyons—startling in the best sense of the word. A landscape that seems conjured by wizards.

I've been on the road for three days now, drifting south from Montana on a loose toss of quiet highways—past the Tobacco Roots and the Ruby Range, the Tetons and the Wasatch, nudging my way toward New Mexico—the starting point for what will be a monthlong trip. I was hoping yesterday to get farther down the road. But the sky was so clean, the mesas so full of shimmer, that getting to this slice of canyon country seemed plenty far enough. Last night I slept outside, under a sky as black as Raven's feathers, shot full of stars.

This morning I'm back to a game I often play in this part of the Southwest. It involves looking out across the red rock hoodoos and twisted canyons from some high place like this one, picking a point twenty or thirty miles distant, and then trying to imagine a walking route to get there. It's ridiculously complicated. Yet no matter which maze of arroyos and canyons I pick, my instinct is to choose a path that, whenever possible, hopscotches between clusters of ponderosa— trees that show themselves easily, what with their soft green canopies shimmering in the sun. I know well how beautiful those open groves would be, their tall, clean cinnamon-colored trunks offering up an almost motherly comfort to the hardscrabble traveler. Especially on searing hot days like this one promises to be, when the ponderosa forest might well be what keeps that sun from eating you alive.

I've come on this trip to witness a vanishing. A human-caused disappearance of these very ponderosa trees—and hundreds of millions more, from New Mexico to California. Starved of water and then taken down by insects and disease, or burned into oblivion by repeated, abnormally intense wildfires, much of what is now ponderosa forest is turning into permanent grasslands and shrublands. Even before the devastating droughts of the early 2000s, fire ecologists declared that the speed at which forests were disappearing in the region to be nearly unprecedented; and, at the same time, that the size of high-severity burns in Southwestern ponderosa pine forests were the worst since the start of the modern climate era, some nine thousand years ago. "Even under the most optimistic estimates of natural regeneration," wrote researchers in *Fire Ecology*, "large high-severity fire patches are likely to remain without forest cover for many decades to centuries."

One study in the prestigious journal *Nature Climate Change* predicted that 72 percent of *all* evergreen forests in the Southwest could die off by 2050, with many more likely to disappear by the end of the century. Here, then, could well be the first post–climate change landscape in America. In 2003, about the time droughts and wildfires were beginning to pile up in the West, Australian philosopher Glenn Albrecht coined the word *solastalgia*. Melding the Latin for "comfort" with the word for "pain," it speaks to what Albrecht saw as a kind of homesickness. A deep unease born out of the sense that one's home landscape is unraveling. This fifteen-hundred-mile journey is an attempt to better understand how such an unraveling came to happen in the American Southwest. What could we have done differently? What can we do now? Also, given the extraordinary aesthetic appeal of these forests—the love affair they've sparked in millions—what will be the shape of the hole in our hearts when they've gone?

I'll begin the search for answers to those questions in the ponderosa strongholds of northern New Mexico and Arizona. Then I'll head off to the northwest, into this canyon country of Utah—a region where the tree can show up either as part of a sprawling village or as a solitary giant in the bottom of some stony canyon. Finally I'll roll west into the legendary forests of the Sierras where, in just the past twenty years, some two hundred million conifers—a great many of them big ponderosa—have already been lost.

Of course when we talk about the forest, we're never really just talking about trees. Disappearing along with the ponderosa are pairs of goshawks, those extraordinary birds found twisting above the canopy in early summer on their legendary sky dances and, while hunting, spinning low around the tree trunks like jet planes in a Pixar movie. Fading, too, is the fluty drumming of Grace's warblers and the hammering of the white-headed woodpecker. The dusky grouse, as well, along with great horned and flammulated and Mexican spotted owls. And then there's the red-tailed hawk and the eagle, sitting in the tops of the big ponderosa, in their nests made of broken sticks. Ebbing, too, are the midnight yelps of the red fox, the chattering of tassel-eared squirrels, the trills of bobcats, the grumbles of black bears. Even dead standing ponderosa have long been vital to the community, what with the holes and hollows in their trunks being prized by everything from bluebirds to nuthatches, flickers to woodpeckers. The Southwest without ponderosa is like an orchestra stripped down to a pair of violins and a kettle drum.

∽

"Its dry and spacious groves invite you to camp among them," wrote the great naturalist Donald Culross Peattie about the ponderosa seventy years ago. "Its shade is never too thin and never too dense. Its great boles and boughs frame many of the grandest views of snow-capped cones, nostalgic mesas, and all that bring the world to the West's wide door." Much like the redwood and sequoia groves of California, mature ponderosa groves sparked awe in both poet and logician. Not just because of the towering stature of individual trees, each wrapped in russet-colored bark and exuding vanilla- or butterscotch-scented perfumes. But also because, in maturity, they form remarkably spacious forests. What's more, big trees rarely wear branches along the first twenty or thirty feet of their trunks; that yields a woodland with sprawling views, the trees fading in the distance into a soft wash of shadows. These aren't the old woods of befuddlement, with waist-deep understories of bugs and brambles—the kind of place Red Riding Hood goes afoul, where the fiends of fairy tales wait for hapless children. Quite the opposite. Here is a forest full of welcome, one that puts its arms around you and lets you breathe. The tree equivalent of a big yellow lab waiting on the front porch, eager to welcome you home at the end of a long day.

Some anthropologists have pointed out that much of the landscape art we've tended to fall for over the centuries tends to mimic qualities of those places where hominins first started walking exclusively on two legs—open, well-lit clusters of trees rising along aprons of grass. The ponderosa is that forest. Maybe the fondness we feel lies deep in the bones.

There are two culprits behind the disappearance of ponderosa in the Southwest. One is climate change. The heat and prolonged drought that have been smothering these lands over the past thirty years can either kill trees outright or, more commonly, weaken them so much they become easy prey for tree-boring insects and mistletoe and blight.

The other culprit, though, has to do with a disastrous choice we made early in the twentieth century—and then leaned into for nearly seventy years—which was to suppress all wildfires. There's much to say about how essential fire is to the forest. For now, though, suffice it to say that in the arid West, fallen trees and branches and other natural debris are cleared away only by wildfire. Put all those fires out, and the debris—often referred to as the fuel load—gets thicker and deeper, until what would've been a healthy, necessary burn becomes a hellish conflagration. Today there are about three hundred million acres of land in the West with unnaturally heavy fuel loads, which is about three times the size of California.

Even under the most favorable climate conditions, the arid Southwest has for thousands of years been a sparsely decorated landscape, showing little of the tree diversity found in the eastern United States or the Pacific Northwest. In the early twentieth century, back in the woods of the East, it was heartbreaking to watch the mighty chestnut fall to blight and devastating, not long after that, to see the collapse of the American elm. But even with those losses, the birds and bears and bugs—and we humans, too—could, each in our own way, keep feasting, literally and figuratively, on the white oaks and red oaks and sweet birch and black locust and sycamore and beech and white ash and tulip trees and hemlocks and white pines.

These western lands have no such bench. Most large pines—and the West is largely a land of pines—live in places with more moisture. In many drier stretches of the region, only the ponderosa has figured out how to thrive to the point of becoming a towering presence. It has an amazing ability to put down deep tap roots in its first year of life, up to two feet long, touching what little water there may be; later, as an adult, it can reach down thirty feet. Add to that its uncanny ability to conserve water by closing the pores of its needles (the tree is roughly four times more water-efficient than a Douglas-fir), as well as the superpower of seedlings able to withstand temperatures of over 150 degrees, and you end up with a tree—the one tree—able to live the bravura of a truly vertical life.

∼

For a lot of beings on planet Earth, these are excruciating times. We're increasingly being knocked about in a landslide of drought and fire and floods and hurricanes and rising sea levels, which even under the best scenario—and at the moment best scenarios have been yanked off the table—will keep us reeling for decades, if not centuries, to come. Like a lot of people, in more anxious moments, I find myself just wanting to escape, forget it all for a while. But sooner or later, I always come back to the urge to pick even one of these thousand doors of loss, pull it open, and walk through. It's not a move without sadness and grief. But embracing the things that are going away also brings riches—the kind of gift that always comes when we stop to honor what we care for most.

This particular doorway is an especially personal one. From the time I was twenty until well into my sixties, the research I did as a conservation writer brought opportunities to spend thousands of miles hiking up remote trails under a backpack—or sometimes, crossing landscapes bereft of any trails at all. It was often the ponderosa that sheltered me when hard weather hit the fan, offering places to hunker down in hurling wind or rain or hail or snow. It offered precious shade, too, on blistering summer afternoons at the bottom of Hells Canyon, or the Chisos Mountains of west Texas, or the sunbaked rocks of Nevada's Spring Mountains.

Yet beyond all that, it was to the ponderosa groves that I went to mourn the biggest losses of my life. My father dying on a construction site when I was twenty-four. My mother, from cancer, when I was thirty-one. My first wife, lost to drowning in a cold Canadian river two weeks past her fiftieth birthday. These were the forests that showed me what it means to have nature hold you. Offering much-needed peace, to be sure. But beyond peace, a feeling of being sheltered—a sense that even when the world is dark, there are places you can go to feel some measure of belonging. Plenty of landscapes have thrilled me over the years: the summits of big mountains; the fast, roiling bellies of Arctic rivers. But when the matter at hand was to salvage myself, nothing suited me like ponderosa.

There was a famous writer born in my hometown by the name of Kenneth Rexroth; often he's been referred to as the father of the Beat poets. Living in the Midwest into his teen years, Rexroth once wrote about waking up one morning and feeling utterly forlorn. For months he'd been out scouring the landscape, hoping to find some kind of inspiration he could use to weave new, more dynamic stories for his generation—something to match "the way of the gods and goddesses and heroes and demigods of the ancient world." But look as he might, out there among the factories and cornfields, he couldn't find it. With the buzz and whim of such myth gone from our lives, Rexroth warned, we'd mostly be left with a gnawing hunger to consume. It was time for Americans to reinvent themselves, he concluded. And that would take reimagining our relationship with nature.

Rexroth eventually packed his bags and headed west, spending his first summer sitting alone in a fire tower above the ponderosa forests of western Montana, writing poetry and watching for smoke. When I was twenty, I went west, too, as I'd told my parents I would when I was nine years old. West to the magnificent mountains, to the whitewater, to the big runs of Douglas-fir and lodgepole pine and ponderosa.

Today when I see the skeletons of the former ponderosa forests—a smattering of bare gray trunks, starved of water or killed by pine beetles, the few remaining green ones standing on point like exhausted soldiers holding vigil for fallen comrades, deep down I feel a kind of primal flinch. It brings to mind a line by poet Jane Hirschfield, written years ago to mark the death of one of her favorite old trees.

"Today, for some, a universe has vanished."

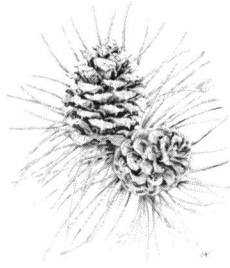

Chapter Two

It's just past dawn, midsummer 12,000 BCE, on land that will one day be called New Mexico. Like a lot of summer mornings in this last gasp of the ice age, this one still pulls with it a restless north wind, tussling the bunchgrasses, sending loud whispers through the dusky five-inch needles of the ponderosa forest. The high temperature on this July day will be just under 50 degrees Fahrenheit—some forty degrees cooler than the average in our own time. These particular trees—combined with a few other scattered groves in the Southwest, and a few near the Pacific Ocean—are the only ponderosa left inside the borders of what will one day become the United States. The offspring of these grandmothers will one day drift up the continent and make homes in much of the West, leading to the ponderosa becoming the most widespread pine of them all. Not quite yet, though. For now, these relatively few trees mostly just hold on. But then holding on is something ponderosa does extremely well.

Every day in this older, colder time, a parade of creatures drifts past the edges of the trees: big dire wolves loping through the grass, pausing now and then to peer into the shadows of the woods. Seven-hundred-pound sabretooth cats. Ten-foot-long, two-thousand-pound giant sloths. There are mastodons and bison and camels too. And mammoths, nine feet at the shoulder and ten tons—about the weight of an army tank—juddering the ground as they pass.

And another animal, too—one that arrived more recently. Sometimes you can find them among these very trees, while at other times they move across the landscape in groups of twenty or thirty, out in the windblown open. They're neither as strong nor as fast as most other mammals. But they're curious. They're clever. And much to their benefit, as omnivores, there's a long list of foods that can keep them going. Now and then they manage to take fish from nearby streams and lakes. Other days they form small clutches of four or five, each carrying a six-foot-long spear tipped with an astonishingly sharp blade fashioned from chert or obsidian. They have what can only be described as a stunning ability to cooperate with each other. And in much the same way as their neighbors the wolves, living with that kind of cooperation is what allows them to now and then take animals way bigger than they are.

But fishing can be spotty, hunting fickle. Even those faster, stronger wolves nearby only manage at best to take large prey about once in every five attempts. So the humans fortify themselves with gifts of the ground, gathering lamb's-quarter and amaranth and purslane. At some point, they also come to prize the inner bark of ponderosa, which as it happens, is packed with carbohydrates, iron, zinc, and vitamin C. Even fats and proteins. The fibrous layers where the cambium cells grow they eat raw, mostly in the spring when the flowing sap turns it sweet. Later they'll figure out how to dry and make flour from that cambium layer, which they'll in turn use to bake bread. Such knowledge will be passed on across hundreds of generations. Today, north of here in Colorado's Great Sand Dunes National Park, fully seventy ponderosa trees in a stand of two hundred bear the marks of Ute people peeling strips of bark in the 1800s.

And then there's the bounty of late summer and autumn. Just like no end of birds and mice and other mammals are doing, the humans head into the forest to gather up what can only be described as a superfood: the oily seeds held in the ponderosa's cones. They're a veritable gold mine of carbohydrates and fiber and precious fats; also magnesium and vitamin E and zinc and manganese and phosphorous.

The longer the people were here, the deeper their kinship with the ponderosa became, and the more the tree gave them. In time they were crafting cradleboards and prayer sticks, baskets and boats and sometimes even snowshoes. They extracted color pigments from the bark for dying cloth and, eventually, for use in Navajo sand paintings. The tree's sap was used as a disinfectant—something the birds had figured out long before, some lining their nests with needles to minimize the risk of infections in the nest. The humans made root medicines for arthritis and back pain, too, as well as teas for aiding digestion. Many still do today.

Beyond that, though, ponderosa was an essential part of some of the first human architecture on the continent. At first, smaller trees were used to hold the roofs of pit houses and sacred kivas. Then, starting around 750 AD and continuing for centuries, the tree became the literal backbone for the multistoried villages of the Ancestral Pueblo. Ponderosa, in other words, gave rise to the first cities in what would later become the United States.

~

In fits and starts, the ice age dwindled, and the region began to dry and warm. Out beyond the edge of the forest, where hickory trees had long grown, the land became increasingly marked by rumples of rolling grasslands dappled with oak and juniper. By then, the ponderosa had started to travel. With every passing year, the years becoming decades and then centuries and then millennia, seeds were plucked from the cones by northbound birds like nuthatches and finches and chickadees. And other birds joined the ponderosa entourage, too, all being especially fond of a tree that produced nuts bigger than those of most other conifers. Some of those seeds dropped from the birds' beaks to land on friendly soil, where they sprouted into young trees. Still other seeds were picked up by humans who, on occasion, carried them across entire mountain ranges, landing them in places where they might not have ended up on their own. It was a patient business, usually measured out just a few miles at a time. And even that sometimes took years.

The descendants of the ponderosa trees in New Mexico and Arizona ran north along the edges of the Rockies, sprouted here and there along the western edge of the Great Plains, then drifted on into Canada. Meanwhile, sister ponderosa trees in the far west, near the Pacific Ocean, made a similarly slow walk up the Sierras, then continued along the Cascades into British Columbia; some also headed inland, toward the Continental Divide. This more westerly variety of the tree became slightly different from those in the east, fashioning their branches with bundles of three needles instead of two. While those in the drier lands of the east devoted themselves to conserving water, their California kin decided to spend more water for the chance at faster growth. It took more than ten thousand years for the two varieties of ponderosa to meet, old friends with shared ancestors, finding each other at last in the mountains of western Montana.

The ponderosa has shown an amazing knack for fine-tuning itself to changing conditions and new opportunities. It can withstand big droughts and searing temperatures in the south and, at the same time, manage deep snow and 30 below zero in the north. Sometimes it even manages to set up shop in wet soils, including along certain floodplain rivers of the Northwest—not typical behavior for pines. Accommodating these different growing conditions takes sophisticated adjustment, fashioning unique traits that are then passed on to future generations. This is why, following a wildfire, it's always best to replant the forest with seeds from trees close to the burn.

Today ponderosa is the primary anchor for no fewer than three dozen unique plant communities, each with its own mix of residents. Here in the drier, warmer low elevations of the Southwest, the tree keeps company with scrub and white oaks, manzanita, blue grama grass, sage, and rabbitbrush. Go a little higher up—keeping in mind that going up means adding moisture—and the blue grama grass yields to fescue and mountain muhly and other bunchgrasses and geraniums, as well as copses of shrubs like locust and snowberry, spirea and serviceberry and fire willow. As moisture increases, other trees join in, too, including Douglas-fir and white fir, Southwestern white pine and blue spruce, even aspen. But the constant is ponderosa. Only in the highest, coldest places are they absent, there yielding the stage to Douglas-fir, white fir, blue spruce, and in some places, up on the very top of the world, bristlecone pine.

~

Of the roughly five billion organisms likely to have arisen on this planet since life began, well over 90 percent have disappeared. Humans may have been responsible for about two hundred thousand of those disappearances, though today, with climate change and habitat loss, that number is growing fast. Those beings that are with us here and now, the plants and insects and animals and birds, are nothing less than world-class champions at the game of life. And among those champions, few have a longer, stronger track record of thriving than the conifers. There are only about two dozen kinds of terrestrial life with individuals capable of living beyond a thousand years. Most are conifers. And while ponderosa rarely reach that age (though a few do), in the right conditions they can routinely make it to four and five hundred years.

The trees in the Pinus genus, which includes ponderosa, came online about 130 million years ago. For a while these so-called gymnosperms—literally meaning "naked seed," referring to how seeds develop on the exposed scales of cones—were able to lay claim to a big slice of the planet. But then came a botanical tidal wave of flowering plants—angiosperms, which means "vessel," a reference to how the seeds develop within the ovaries of flowers protected by fruit. It was in large part due to the overwhelming success of flowering plants that the pines, a bit back on their heels, started coming up with new strategies for staking out less crowded places to call home. Trees like the ponderosa made the marginal into the marvelous, adapting to thrive in more stressful habitats than most flower-bearing plants could manage. In particular, the ponderosa did well overcoming two especially challenging landscapes: those with poorer, drier soils and those frequently touched by fire.

For all the clever tricks ponderosa has up its trunk, one essential part of its life here in the Southwest has to me always seemed something of a tightwire act. For each next generation of ponderosa to get their best chance in the sun, in addition to the modest seed crops it produces in most years, some years it goes all in to produce way more than that—a so-called mast crop. How often the trees manage to do this depends on climate conditions, but on average it's about every four years. The first thing that needs to happen is a warm, wet spring, which—two and a half years later—will lead to a massive crop of seeds in the fall. But that's just the beginning. In addition, when those seeds drop from the cones and hit the ground to germinate, they ideally need to land in moist soil. And that holds true as well for their growth as saplings the following spring: There needs to be an above average amount of moisture. Many of the Southwest's ponderosa live in very dry places—lands with precipitation levels barely above an actual desert. Asking for a wet spring, and then a wet autumn two and a half years later, and then another wet period the following spring seems like a lot. Then again, I'm writing from a somewhat parched perspective, given that I'm living today in the thirty-first consecutive year of a Southwestern drought. In earlier times, when the ponderosa was setting up these kinds of alignments, it was doing so amid far friendlier climate patterns.

While in a smaller seed production year a ponderosa may lose up to 90 percent of its seeds to hungry birds and other animals, with the overwhelming number produced in mast years, those losses drop to about 10 percent. Note that it isn't just one or two trees in a grove that will go about making these bumper crops. Those efforts usually involve an entire grove, the trees coordinating with one another to all produce their pollen at the same time, sharing it with one another, thereby promoting genetic mixing. And that genetic mixing increases the overall strength and resilience of the forest.

Pines tend to have waxy coatings on their needles to help retain moisture, which of course comes in handy in the arid West. What's more, those needles are thin and more or less flat, which means big winds can roar through without breaking branches. Pines in colder climates have also figured out a way to concoct what amounts to antifreeze. This amazing chemical brew causes ice to form in hexagon-shaped crystals instead of the usual sharp needle shapes, which keeps the cells from being pierced and damaged. And then there's the blessing of being evergreen. When deciduous trees drop their leaves in the fall, they lose access not just to photosynthesis, which pines can conduct to an extent even in winter, but to the nitrogen, potassium, calcium, and phosphorus stored in their leaves. Be a pine tree, though, and by hanging on to your needles, you'll also be able to hold on to essential nutrients.

For all these brilliant strategies, the one that really speaks to me on this particular journey is the fact that ponderosa have figured out how to make friends with fire. The signs of that relationship are everywhere, from the base of the trees' trunks to the tips of their highest branches. The whimsical orange-and-rust-colored jigsaw pieces of the bark, for example, are designed in part to slough off if they catch fire, shedding flames to the ground rather than letting them run up the tree and burn the crown. Also, as mentioned earlier, the reason we get such a deliciously spacious feeling in a mature ponderosa forest is partly thanks to there being no branches on the lower trunks. As the trees drop their bottom branches—the ones that get less sun—it has the effect of keeping flames from climbing branch to branch into the crown. What's more, ponderosa needles, long and moist, form a kind of protective feathering around the buds, shielding them from the heat of flames. And if they do happen to get burned off, the tree will get busy making more, and do so in surprisingly little time. A ponderosa tree can lose more than 90 percent of its needles and be back in business with a new set the following year.

In short, this most widely distributed pine in North America achieved that status by figuring out how to thrive in a challenging world. And yet humans, over the last century especially, have pushed the environment past where even a super survivor like this one is struggling.

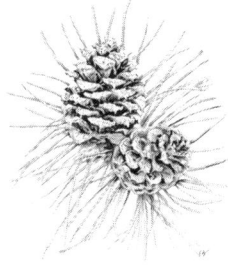

Chapter Three

There's an extra measure of poignancy to the wreckage of climate change when the casualties are trees. Any kind of tree. The ones wrapped in gravitas, like the white oaks of New England and the tulip trees of Tennessee, the California redwoods of Mill Creek and Bull Creek Flats, the big maples of Indiana and Illinois. And those more modest in stature, too: the red alder of Alaska and the Pacific Northwest, or the aspen groves in the southern Wasatch Mountains of Utah and southwest Colorado, where the trees look like ballerinas in pirouette.

Part of our affection for trees arose from a long list of practical gifts they've given us: providing shelter, offering us quick escape from hungry predators, feeding us with fruit and nuts and holding no end of honeycomb, giving us the fire that cooked our food and kept us warm. In more recent times, here in the West it was often the ponderosa we cut down and sawed into coarse planks, nailing them together into jails and rooming houses and long plank bars in shoddy saloons. We used the tree to make the ties that held the railroad tracks and then used still more of them to fire the steam locomotives that ran on them. Ponderosa built the ranchers' homes and barns and irrigation gates, the corrals and the windmills and the pasture fences. They shored up the mines and made our boats—including the ponderosa that Lewis and Clark, guided by the Nez Perce, made into five large canoes they then paddled down the Columbia River to the Pacific. During the Civil War, when the Union Navy ran out of pitch and turpentine to tend their ships and could no longer get it from southern pines, they drove their taps into California's ponderosa trees. Some lone ponderosa, standing tall on the northwestern edge of the Great Plains, became signposts for west-bound settlers in wagon trains—offering not just direction but flashing the hugely welcome news that the open, treeless plains were finally coming to an end.

But our long arboreal love affair goes way beyond the practical. You can see it in the way the ancient Greeks bestowed sacred honors on the oak and the cayuga—much as people living in the British Isles did for the yew. Or the Arikara people of the Great Plains bringing reverence to the cedar, or the Conibo of the Amazon to the catahua. For the Norse it was the ash tree, for the Chumash the pinyon, for the Germans the linden. For others it was pine. Or banyan. Or sycamore. Or fig. For any culture living in or near trees, those trees were sources of the sacred.

Some of the oldest myths in the world say that humans were once trees—or, conversely, that trees were once humans. The Norse creation story describes how the first two people were fashioned from two tree trunks the gods found lying on the beach: From one trunk, that of an elm tree, they created a woman named Embla, and from the other, an ash tree, they created a man named Ask. Meanwhile, two thousand miles away, a creation story from Persia emerged to tell how a tree grew out of the decaying body of the first human. When the trunk of that growing tree split into two forks, one became man and the other woman; from the fruits of the tree came the different races of humans.

German folktales speak of babies starting out as spirits in the underworld and then passing through trees to become human. In similar fashion, people in the British Isles said each of us had a parallel soul, a double, which lived in a nearby tree. In early America, farmers from the Atlantic Seaboard to the Midwest planted trees on the arrival of a newborn, believing that as the tree grew strong, so would the child.

On perhaps an even more ethereal note, cultures around the world have also treated trees as a link between Earth and the heavens, allowing us to communicate with the divine. The Bible mentions trees over five hundred times—more than any other topic besides humans. The ancient Jews lit torches made from the branches of the oil pine—the tree of redemption. The first temples in ancient Gaul were sacred forest groves, while in Japan Shinto shrines are to this day cradled by sacred forests. The Aztec spoke of trees that lifted up the sky.

For Christians olive trees symbolized health and fertility, while for Muslims they meant renewal. In Islam, on Ramadan—a celebration of compassion and generosity—people planted trees, as they still do today; when those planted trees later produce fruit or nuts for a bird or an animal or a human, the action is charted in the heavens as an act of charity. Egyptians sought Osiris in the branches of a sycamore. Buddha was shown the four great truths in the middle of a forest. And it was while standing in a grove of oak trees at Mamre that Abraham rubbed elbows with God.

In colonial America, images of trees were sewn into state flags, stamped into coins, stitched into quilts. When the Civil War broke out, some suggested we confess our moral decline by cutting down the American elm that served as our national tree. One hundred and forty years later, when a modest pear tree growing at the base of the former World Trade Center managed by no small miracle to survive the attacks of September 11, we nursed it back to health and made it sacred. In the years since those attacks, everyone from stockbrokers to dockworkers, lawyers to line cooks, prime ministers to presidents of the United States have placed wreaths against its trunk.

Writer John Fowles speaks eloquently about these more transcendent, more inscrutable relationships we have with trees. "If I cherish trees beyond all personal need and liking of them, it is because of this, their natural correspondence with the greener, more mysterious processes of mind; and because they also seem to me the best, most revealing messengers to us from all nature, the nearest its heart."

~

On one hand, then, the disappearance of the ponderosa will leave us without many of the practical gifts still being gathered today: using the wood of the tree for joists and rafters, siding and decking, cabinets and window frames; boxes and crates, furniture and fence pickets and firewood and toys. And other deeply functional gifts, too, like how the ponderosa forest holds snowpack in place and keeps it from evaporating in the direct sun—incredibly important, considering that snowmelt provides well over half of all flowing fresh water in the West.

But beyond all such "services," if you will, the disappearance of the ponderosa forests will impact as well our myth and story, weakening the kind of natural correspondence we have with trees that Fowles talks about. All this might seem squishy and imprecise, barely worth mentioning in the midst of our ongoing environmental disasters. But squishy and imprecise is the nature of caring about things. And no matter how brilliant our technological prowess may be, our future is going to depend on how deeply we can fall in love.

That sort of love, that kinship, has a lot to do with why this morning I'm fifteen miles north of Taos, New Mexico, roaming on foot across a small slice of rural land owned by the University of New Mexico, known as the Lawrence Ranch. To the west runs a loose toss of ponderosa forest, in five miles yielding to a stretch of sage and pinyon hills tumbling to the edge of a set of sheer cliffs above the Rio Grande. Meanwhile, to the east are the Sangre de Cristo Mountains—a 250-mile-long swell of heartbreaking beauty, running from Poncha Pass in Colorado all the way to Santa Fe. The Lawrence Ranch is the kind of place you might see on television some evening, then later find yourself daydreaming of quitting your job and moving west. For at least a thousand years, this land was home to the residents of Taos Pueblo.

In the early 1800s it was also used intermittently by the Kiowa people, leading to it later becoming known as the Kiowa Ranch. In the 1880s the land was homesteaded by John and Louise Craig; after them came a woman who raised Angora goats. Then, a century ago, in 1920, the Kiowa Ranch came into the hands of a fantastic, fiery force of nature named Mabel Dodge Lujan. And Mabel would have much to do with helping the world fall in love with this landscape, including with its signature tree, the ponderosa pine.

Mabel was a wealthy banking heiress, a prominent New York socialite, and also a popular, outspoken syndicated columnist for the *New York Journal*. In 1917, to the shock of her friends, she abandoned Greenwich Village and headed off to live in New Mexico. She explained her abrupt and to many inconceivable decision simply by saying she needed a change. As a follow-up years later, she added, "And I got it."

Even as a young woman, Mabel was appalled by the "ghastly social structure" of the wealthy, including her own family, not least because of the stifling limits the system put on women. She pushed hard and often for "a world where a woman could choose her own role in life." After a serious romantic partner one day just up and left without warning, breaking things off with a hastily written note, she lamented to the readers of her newspaper column that she was "tired of being a mother to men." Soon afterward she got a letter from the man who'd been her third husband, who at the time happened to be traveling around Santa Fe. He told Mabel of the strong Native cultures he'd found in northern New Mexico, saying there was much to learn from them. "You could let the American people know," he told her, "that there are other forms of civilization besides ours." Mabel thought it was a good idea—one that seemed to be appearing at just the right time.

Once settled in Taos, she wasted no time tapping into her network of luminary friends around the globe—poets and painters, photographers and free thinkers of every conceivable persuasion, inviting them to gatherings at her New Mexico home and salon, which she called Los Gallos (the roosters). Before long she'd created a rustic version of the salons she'd held in Greenwich Village—and before that, in Paris, with her good friend Gertrude Stein. She gave a bolt of energy to the fledgling Taos Society of Artists, helping to turn it into a major player in the growing avant-garde cultural scene. A scene taking much of its breath from the northern New Mexico landscape—a place that Mabel, on first seeing it, described as "strange and terrible and sweet."

In 1923 she married her fourth husband, a Tewa Indian from Taos Pueblo named Tony Luhan—a man, she said, with "the sun in his heart." Tony, too, would become a key player in the emerging art scene, making it possible, among many other things, for Ansel Adams to create his first book, *Taos Pueblo*. Some disparaged the couple. Some of the criticism came from Tony's own people, who accused him of abandoning the traditional way of life; at the same time, Mabel was admonished by some in her own culture, who frowned on the mixing of races. No matter. Mabel and Tony liked each other. Beyond that, Tony taught Mabel a great deal about his culture—a culture she found considerably kinder and more just than her own.

Mabel would help northern New Mexico become host to one of the most impressive art movements in American history—a renaissance of painting and pottery and photography and writing, courtesy of everyone from D. H. Lawrence to Georgia O'Keeffe, Gustave Baumann to Barbara Latham, Andrew Dasburg to Ansel Adams. The vast majority of it would be shaped in no small part by what they saw in the nature around them: the Sangre de Cristo Mountains, of course, as well as Miranda Canyon and Tent Rocks; the Rio Grande Gorge, the Red River, and the Rio Pueblo. They came here and saw sunlight pouring through some of the cleanest skies imaginable—and on summer afternoons, watched towering cumulous clouds let loose with startling shows of thunder and lightning. They walked in air scented with the clean bitters of sage and pinyon. And on many days they stopped for a moment to relish the ponderosa, the trees standing up and standing out, softening the sun, feathering the distant ridges.

As she had a habit of doing, Mabel became infatuated with D. H. Lawrence before ever meeting him. She admired how he'd been willing to buck the status quo—by railing against the dehumanizing effects of industrialism but also by jolting stale Victorian morals, something he later did in a big way with his often-reviled *Lady Chatterley's Lover.* (That work, which one critic described as "the book that brought good sex writing to the masses," wasn't allowed to be published in the US for thirty years.) But of all Lawrence's talents and persuasions, most important to Mabel was his ability to render place. On reading his travel narrative *Sea and Sardinia,* she became convinced no author would be better able to capture the beauty of the New Mexico landscape and culture. So she sat down and wrote a letter, inviting Lawrence and his wife Frieda to Taos.

They arrived in 1921, staying at first in an old cabin near Mabel's home and salon. But soon Mabel set them up several miles away, in a simple writing cabin on the Kiowa Ranch. The writing cottage was small and humble—just an old homestead cabin from 1891 built out of ponderosa pine, to which Frieda and "Bert," as Mabel called Lawrence, set about adding two small front porches and an adobe brick fireplace. Mabel later gave the place to the Lawrences, asking as her only payment the manuscript to Lawrence's new novel, *Sons and Lovers.*

Towering just outside the door of the cabin, enthralling Lawrence from the very first time he set eyes on it, was an altogether marvelous lone ponderosa.

"Here on this little ranch under the Rocky Mountains," he wrote, "a big pine tree rises like a guardian spirit in front of the cabin where we live . . . Standing still and unconcerned and alive." Lawrence liked to write from a wooden bench and table at the base of the tree, where he crafted several poems and essays, as well as parts of *The Plumed Serpent* and his novella, *St. Mawr.* Sometimes, in the middle of the night, he'd go out and lie on his back on that bench, watching the stars winking through the branches. "The tree has its own aura of life. And in winter the snow slips off it, and in June it sprinkles down its little catkin like pollen tips, and it hisses in the wind, and it makes a silence within a silence."

Mabel had been right in thinking D. H. Lawrence would be a perfect author to write about this place. "I think New Mexico was the greatest experience from the outside world that I ever had," he later confessed. "It changed me forever." Years after his death, Frieda had his remains cremated; the ashes are interred here at Kiowa Ranch.

The couple was on the ranch for five months in 1924, then returned in 1925, when Lawrence was thirty-nine. And he had every intention of coming back yet again. But his fading health never allowed it. "It grieves me to leave my horses, and my cow Susan, and the cat Timsy Wemyss," he noted before departing in 1925. It grieved him, too, to leave the big ponderosa outside the front door of his cabin. Lawrence found spiritual anchoring in northern New Mexico. He likened Taos to the monasteries that arose in Europe during the dark ages—places that, while vulnerable, were "never overcome in a world flooded with devastation." They alone kept the human spirit from disintegration. A person coming to Taos would sense something similar, he said. Here there was an unmistakable feeling of arrival.

One of Lawrence's American literary critics, writing about the same time the author was hanging out in New Mexico, commented on his uncanny ability to "enter" whatever thing he happened to be looking at. One of the best places to feel that ability is in the passages he wrote about this ponderosa. "In the over branching darkness . . . one suddenly realizes that the tree is asserting itself as much as I am. Our two lives meet and cross one another." The energy of the ponderosa, he said, left him "more bristling and turpentiny." At the same time he imagined the ponderosa took from him "a certain shade and alertness." Lawrence claimed that if he wrote off such thoughts as ridiculous, seeing the tree only in terms of how old it was or how much lumber it held, as so many people would, then he would be forsaking a part of himself. You can either shut the doors of perception, he argued, or open the doors that have already been shut. "For man, as for flower and beast and bird, the supreme triumph is to be most vividly, most perfectly alive."

Now, on this sunny summer day a century later, here I am cozied right up to Lawrence's beloved tree. It has a kind of unfussed regalness. The thick furrowed trunk runs clean for the first thirty feet and then pops into loose tiers of stout branches. Some of those branches run straight out and others meander, each on its own journey to find the sun. Taken altogether they create a pose that feels nonchalantly happy. I wrap my arms around the trunk. Then, with my head tilted back and my cheek against the bark, I stare up through those whorls of shiny green needles into a flawless New Mexico sky.

When I first arrived at the ranch this morning, eleven high school students from Albuquerque were finishing up a summer class visit. English students, the ranch caretaker informs me. In truth they looked a whole lot more satisfied than I would've thought possible from a visit to the writing retreat of a white guy dead for some hundred years. The secret may have been their teacher: an animated, thirtysomething woman—intense and friendly—helping them map some of D. H. Lawrence's impressions onto their own lives.

"What about that question we talked about earlier?" she asked. "What do you think Lawrence meant when he said that in New Mexico time is different?"

I caught only one answer—from a tall, awkward-looking boy wearing pale blue glasses.

"He always liked whatever nature he found walking in England. You know—flowers, trees, all that. But nature's bigger here. So big it got into his head, busted his whole way of seeing things."

After taking in the tree a while longer, at one point catching a red-tailed hawk float to a rest on one of the upper branches, I walk over to the front window of the writing cottage; on the other side of the old, wavy glass is a spare and colorless little room with a small writing desk. I can almost see Lawrence sitting there, a tiny smile of satisfaction on his face at some clever line he'd captured. Or maybe he's scowling, having had yet another row with Mabel; this time angry about her constant evangelizing about that new thing called psycho-analysis, which he considered—to use the English vernacular of the day—so much codswallop.

~

D. H. Lawrence wasn't the only artist to find himself making friends with this particular ponderosa. Among the others was Georgia O'Keeffe. Like Lawrence, she also found her way to the Kiowa Ranch by way of an invitation from Mabel Dodge Luhan. And as had happened with Lawrence, O'Keeffe fell head over heels for northern New Mexico.

By the time she set foot here, O'Keeffe had been a fan of Lawrence for more than a decade. She loved how his writing elevated the sensuous—not as some kind of cheap sensationalism, but as a means of exploring the edges of his creativity. She was a fan, too, of how he celebrated nature, plumbing relationships in ways that to her seemed deeply spiritual. This quality O'Keeffe was especially drawn to, leaning on it for her own creations, paying homage with brush and canvas to the transcendent qualities of the natural world.

Taking up residence in the Lawrence cabin, she immediately found herself smitten by the old ponderosa. One still and perfect night, she went out to the same carpenter's bench Lawrence had used, lying on her back as he had, looking up at the stars shining through the branches. The result was an extraordinary painting—a lively, poignant modernist joining of space and time, earth and sky, described by one critic as an homage to the wholeness of heaven and earth. Though O'Keeffe wasn't in the habit of naming her paintings, she went from first calling the work *Pine Tree With Stars at Bretts, N.M.*, to, on learning of Lawrence's death, *Lawrence Pine Tree, With Stars*, later still to *The Lawrence Tree, Night*, and finally to the name used today, *The Lawrence Tree*.

The mountains, the hills, the light, the sage-scented air, the wild-flowers, the pines—for so many people who came here, it all seemed to mingle together and seep into the heart. And no one who felt it happen would ever forget.

"In a cold like this," wrote D. H. Lawrence shortly before leaving New Mexico for the last time, "the stars snap like distant coyotes, beyond the moon. And the pine trees make little noises, sudden and stealthy, as if they were walking about. That place, the ranch, heaves with ghosts. But when one has got used to one's own home-ghosts . . . they are like one's own family, but nearer than the blood. It is the ghosts one misses most, the ghosts there, of the Rocky Mountains, that never go beyond the timber and that linger, like the animals, round the water-spring. I know them, they know me: we go well together."

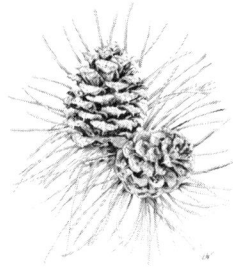

Chapter Four

From the Kiowa Ranch I head southwest, soon reaching a heart-soothing sprawl of middle-of-nowhere New Mexico, finally coming to a stop at the bottom of a vast volcanic crater known as Valles Caldera. The belly of the crater is stitched in wheatgrass and grama and dropseed and squirreltail, spiced here and there with wild iris and monkey flower and sedge. But best of all is the 125-acre community of old-growth ponderosa known as the History Grove. Most of the trees here are youngish elders, about four hundred years old, which means they were saplings about the time Shakespeare was conjuring *Hamlet*, and the *Mayflower* was nosing into the shore of what would become Massachusetts. Given the ferocious cutting of ponderosa in the nineteenth and twentieth centuries, which by 1950 had wiped out the vast majority of America's mature groves, it was either dumb luck or divine intervention that allowed the History Grove to escape a similar fate. Some say the trees are standing mostly because a wealthy tool supplier named James Patrick Dunigan, who bought the land back in 1963 when it was still the historic Baca Ranch, wasn't keen to sit on his nearby porch and look at stumps.

Whatever the reason, I'm more than a little grateful they were spared. It's a thrilling, thoroughly inspiring forest—maybe something on the order of what Claude Monet might have seen in the old-growth forests of Fontainebleau, or Canadian artist Emily Carr found in British Columbia. The exceptionally big, clean lines of the trees here also speak of architecture, a good reminder of how so many of the world's great buildings were designed to honor the heft and grace of big trees. The Parthenon, with its grand columns paying homage to the sacred groves of Greece. Saint Peter's Basilica in Vatican City, its uprights mimicking trees reaching for the heavens. Ta Prohm temple in Cambodia. The Great Mosque of Djenné. The Temple of Apollo. The Alhambra.

My visit starts in the best way possible, with a slow, aimless saunter from one elder to another. I pause for a long time at one of the biggest trees, pressing my nose into her bark, breathing in her vanilla and butterscotch. I've never encountered a tree that smells more delicious than ponderosa—a scent arising from chemicals called terpenes, which are produced in the wood and resin and needles. Terpenes are actually a part of both the immune and defense systems of many plants, and the source of much of what so delights our noses: rosemary and lavender, clove and dill and cardamom, cinnamon and basil and thyme, parsley, dill, and oregano.

The terpene game as the ponderosa plays it is a sophisticated one, with the specific formula tweaked and remixed as needed. For example, whenever the number of birds feeding on a tree grows past a certain point, it serves as a cue for the tree to muster energy and change the terpene flavor of its bark, making it into something less palatable to insects. As if the tree "knows" too many birds means too many bugs. Likewise, in the event of a fungal infection, the ponderosa can send a special terpene brew into the current growth ring, essentially creating a barrier that lessens the risk of the contagion spreading.

There are no other people in the grove this morning. Though the morning light is building fast, it's still soft, filtering through a steely green weave of needles some hundred feet in the air, finally touching down on the ground around me in something more candle than sun. A goshawk shoots past overhead, sending a pair of Abert's squirrels, who are a bit prone to apoplexy anyway, into fits of twitching and clucking. The bird is likely fresh off the nest after a month sitting with her young, free again to join her mate as part of one of the fastest, most graceful hunting couples in all the bird world. Five minutes after the hawk disappears, the squirrels are still barking; when they finally calm themselves, it's quiet again. A pair of chickadees are doing the Lindy Hop on one of the lower branches without saying a word—happy when they're plucking ants from between the crevices in the bark, and happy when they're not.

Several of the trees in this grove wear good-sized fire scars—horseshoe-shaped burn injuries that the tree will keep for the rest of its life. When a ponderosa is injured like this, it exudes a layer of pitch—a kind of preservative incredibly rich in antimicrobials—thereby sealing the wound. One reason why people like ponderosa so much for cooking and heating is because normal, frequent wildfires prompted just this kind of pitch release; as a result, the wood burned long and hot. Dendrochronologists also appreciate pitch-rich ponderosa, since big trees that were felled even a century ago may still have intact stumps, and therefore intact tree rings. Tree scientists in future generations might not be so lucky. Due to our having suppressed fire for so long, the pitch response is less frequent, leaving dead trees and stumps more prone to decay.

Wandering on through the grove, I come across one of the younger trees with huddles of ants moving up the plates of bark, herding small groups of aphids like six-legged cowboys rounding up cattle. They'll eventually steer the insects up the tree and out to the branches and twigs, where the aphids will use tree sap to produce that sugary "honeydew" ants can't seem to get enough of. In return, the ants protect the aphids from wasps and ladybugs. On the same tree, just above the level of my head, is a nearly perfect square box of twenty carefully drilled holes in the bark, where a sapsucker tapped the tree to, appropriately, sip the sap. I don't have to look long before spotting him, perched on a high branch in an adjacent tree. The longer and deeper you walk into a ponderosa grove, the more likely you are to find yourself standing before a veritable parade of residents, from birds to butterflies to small mammals. Ponderosa is, in fact, the most species-rich pine forest on the continent—hosting close to a hundred species of birds alone.

Meandering alone back and forth in the History Grove, I'm doubling down on the idea that in all the world, forest walking finds one of its highest expressions in the sumptuous, uncluttered way of a mature ponderosa grove. By the time I leave such places, I'm convinced all over again that for all our sophisticated ways of traveling, it's hard to beat simply being out on foot. As Rebecca Solnit described it, "The rhythm of walking generates a kind of rhythm of thinking."

The extra appeal of being on foot in a mature ponderosa forest, though, is that you really can wander completely aimlessly. This a place to be freed from the scribed lines of trails, allowing a chance to travel more like a child goes through the world: intrigued by one thing over here and another over there, trading linear progress for pointless delight. Curiously, studies have shown that such meandering can help expand the brain's capacity for nonlinear thought—essential to creativity and critical thinking.

For me, though, at least these days, aimless meandering in the woods is mostly about the chance to grow calm by mingling my inside with the outside. To fall back into one of the great luxuries of being human—which is to be in a beautiful place, with nothing to do.

∽

And yet outside the rare slice of old growth like this one, the ponderosa forests we see today, while entrancing, are vastly different from what they were in the late nineteenth and early twentieth centuries. Edward Beale, appointed by Abraham Lincoln to be Surveyor General for the United States, describes well what the average ponderosa forest was like back then. Working to lay out a wagon road in northeast New

Mexico—basically the same road that forty years later became Route 66—Beale was dumbfounded by the ponderosa he encountered. The trees created "the most beautiful region I ever remember to have seen in any part of the world." A vast reach of gigantic pines, "every foot covered with the finest grass, and beautiful broad grassy vales extending in every direction. The forest was perfectly open and unencumbered with brush wood. The travelling was excellent." Beale's fellow army officers often made similar observations, describing mature ponderosa groves as places where you could ride a horse through at full gallop or, more amazing still, make easy passage in a box wagon.

So what happened to those singular, celebrated forests? Again, millions of trees fell to the saw. But even forests that haven't seen logging for decades aren't growing back to anything like their former stature. There are a couple reasons for this, but let's start with this one: The forests we see today were profoundly shaped by the fact that for much of the twentieth century, our management of the western forests was yoked to an ill-advised, hubris-soaked mission to quash all wildfire from the landscape.

On some 500 million acres of public lands in the interior West—the Sierras to the Rockies, the Santa Catalinas to the Sawtooths, the Black Hills to the east slope of the Cascades—regular wildfire does nothing less than drive the cycle of life. The dry climate of the region means far fewer of the sorts of microorganisms that in other, wetter places break down dead plant material in fairly short order, releasing the nutrients it holds back into the soil. In the arid West, though, the primary way dead plants and trees are broken down is through fire. Which is why so many plants here, from fireweed to globe mallow, pine grass to lodgepole pine, evolved to take quick advantage of the nutrients held in the ashes of recent burns. Taking fire out of the ponderosa forest was mistake number one, and the consequences will be following us for years to come.

〜

I walk out of the History Grove and head up a small slope to the west, onto land still wearing scars from an eleven-year-old burn sparked by a downed power line, known as the Thompson Ridge Fire. As fires go, or I should say as fires go these days, at 24,000 acres, it was on the modest side. The effort to fight it included protecting a nearby huddle of historic ranch buildings, as well as the History Grove itself—work initially given to a team of skilled quick-response firefighters from Arizona known as the Granite Mountain Hotshots. The hotshots did their job with their usual proficiency and moved on, passing the rest of the work to an elite regional team known as the New Mexico Returning Heroes, made up of a group of returning military vets.

Two weeks later, those same Arizona hotshots were on a blistering, rapidly burning lightning-caused fire in the drought-stressed chaparral of central Arizona. It was called the Yarnell Hill Fire. Once again the crew did their gritty, lung-busting dance—this time in far more rugged terrain—wielding chainsaws and pick-and-hoe-like tools called Pulaskis, building fire lines at a level of exertion that would leave most of us crumpled in a heap. It was 100 degrees. The kind of conditions some wildland firefighters still call "tits up" weather, referring to how dead cattle are often found on their backs, udders to the sky. Like most catchphrases firefighters toss around, this one wasn't far off the mark. Six years ago, when OSHA was looking into the general health of wildland firefighters, they found nearly all had experienced heat-related headaches, dizziness, nausea, vomiting, or other illnesses. Almost none reported it, figuring it to be part of the job.

On the afternoon of June 30, 2013, pushed by a powerful, erratic shift in the winds, the Yarnell Hill blaze made a dangerous about-face. Suddenly big flames were running straight for the hotshots, out on foot traveling from one "safe zone," where the forest had already burned, to another one farther down the mountain. The fire was galloping too fast even for athletes like these to outrun it. With massive walls of flame roaring down on them, and no time to clear away the vegetation to create a nonflammable safe area, nineteen firefighters dove into their tiny, heat-reflective fire shelters, sealed them up, and prayed for the best. The best didn't happen. A two-thousand-degree wall of flames overran the crew, killing all nineteen.

All of this has me leaving the beautiful History Grove—one of the last places where things went just right for the Granite Mountain Hotshots—in a downhearted mood.

The largely human-caused reasons ponderosa are dying throughout the Southwest—which include overzealous fire suppression—have at the same time left wildland firefighters wrestling with increasingly treacherous burns. It's true that risk has always been a part of this work. But the bigger, hotter burns of today are making an already dangerous profession more so. In that hot, drought-ravaged year when the Granite Mountain Hotshots died, wildland firefighting became the second deadliest profession in America. And the increased demands on firefighters can make for other kinds of losses too. Depression, eating disorders, and alcohol abuse are soaring; in one recent survey, just under 60 percent of respondents said they personally knew a colleague who had died by suicide.

The towns that have been through severe wildfires never, ever fail to pour out plenty of thanks to the firefighters; they know well that whatever was saved was saved thanks to them. Yet for doing this staggeringly hard work, many are earning paychecks smaller than what they'd make working in a ski town Taco Bell. (Meanwhile the prisoners who are increasingly being tapped to fight wildfires—prisoners made up a third of firefighters in the 2025 wildfires in and around Los Angeles—make far less still, under six dollars a day.) In the years to come, more and more responders are going to be needed to deal not just with wildfires but with hurricanes and heat waves and floods and tornadoes. If we're going to expect a heroic response, then we should treat those doing the risky, backbreaking work like heroes.

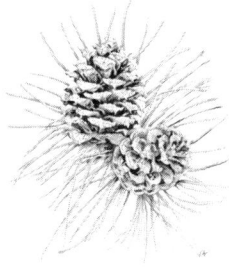

Chapter Five

I can't remember a time when my life wasn't lit by trees. Still in my head are fuzzy movies from when I was three, in South Bend, Indiana, out with my mother scouring the ground under the hardwoods on 27th Street, looking for the red feathers of cardinals. Or out in the shade of one of the big oaks in Potawatomi Park, picking at sky-blue bits of eggshell from a just-hatched robin—a bird my mother said would, incredibly, one day just jump out of the nest and fly. Or sitting alone against the lone young maple tree in our tiny backyard, looking up and out beyond her upper branches, watching animals made of clouds appearing and disappearing in the pale Midwestern sky.

As I got a little older, when school was out for the summer, there were the trees with their arms out, asking me to clamber up and sit for a while. Now and then on random Saturdays, my parents loaded my older brother and me into the back seat of our boxy Chevrolet Impala and drove us out of town, past the brick warehouses and the cornfields, on the lookout for some particularly good stray oak or maple by the side of the road suitable for climbing. I can still see my father standing at the base of the tree, stance squared with his arms held up, squinting through the leaves, cautioning me not to put my weight on any of the dead branches he could see from down below. To this day, for me it remains one of the best dad things ever.

Curiously, the tree that ended up holding so much of my own story is one that back then I'd never met. I first encountered it on Sunday nights in the early 1960s on a black-and-white television—my brother and I lying on the living room floor and eating grilled cheese sandwiches and potato chips, entranced by a simple, absurdly popular horse opera called *Bonanza*.

Filmed on a sprawling ranch east of Lake Tahoe, the show opened with credits rolling over the four smiling Cartwrights astride their horses. Behind the boys was the inlet of a lake, and behind that a rumple of foothills covered in stately ponderosa. As far as I was concerned, the tree was part of the cast: towering behind the Cartwrights as they headed off to rope cattle; a grove rising behind some young schoolteacher driving a springboard wagon to Carson City; as a backdrop for thieves on the run; even as setting for one of the boys out on a rare date. Ben Cartwright even named his ranch The Ponderosa. (But then really, what else could he have called it? Aspen Ranch sounds like a rehab center. The Lodgepole? More suited to men with fezzes in go carts. The Juniper? Somewhere Hoss might go to get exfoliated.)

In 1959, a staggering twenty-five prime-time television Westerns rode in to rule the airwaves. And ponderosa trees—the darlings of many film location scouts—made at least cameo appearances in a lot of them. Through the early 1960s, parents of the Boomers were spending a staggering $100 million a year—about a billion in today's dollars—on cowboy hats, cap pistols, chaps, and other cowboy bling for their kids. My brother and my friends and I swallowed it hook, line, and sinker, envisioning ourselves on some kick-butt ranch—figuring that for us, too, just like for the Cartwrights, the best life was the one lived whistling down the wind on a lonely trail with a good horse.

But it wasn't just us kids who were smitten. In 1963, no less than Johnny Cash penned a special song about *Bonanza*, basically adding lyrics to the television show's instrumental lead-in. Which means there was actually a time back then when you could be driving to your job, stuck in traffic, and suddenly find yourself singing along with Johnny Cash on the AM radio about the mighty Cartwrights—belting out lines like: *Hand in hand, we built this land, the Ponderosa Ranch! Our birthright is this Cartwright bonanza!* And my personal favorite: *Singing pines of boundary lines for the Ponderosa Ranch!*

Also intended to please the adults was the 1964 release of the album *Welcome to the Ponderosa,* by Lorne Greene, who played the show's wise and patient but don't-piss-me-off father. The album took its name from a six-minute talking ballad called "Saga of the Ponderosa," featuring a ranch-sized dose of the Nelson Riddle orchestra to smooth out Greene's messy attempts to stay on key. In the song he talks about being a young, freshly retired sailor on the East Coast when suddenly his young wife dies, leaving his heart empty—and a newborn son named Adam in his arms. Soon after he's beguiled by a mysterious talking wind. It was that wind, he explains in his grave and gravelly baritone, that kept drawing him and his son ever farther west—a journey over several years that brought the birth of two more sons and the death of two more wives. He kept going, though, all the while determined to unravel the meaning behind the one word that the wind kept whispering in his ear: *ponderosa.*

Heroes in Westerns were portrayed as avatars of freedom. But looking back, it was a freedom of monumental exclusion, where God himself had given the land over for the pleasure and profit of white men. Native Americans were too often treated as simple and savage, their only salvation said to lie in adopting the ways of their colonizers. Women characters, too, were fenced in by stereotypes, including the one where they seemed destined to fall in love with men who were free of any need for love at all. What we white boys and our fathers watched unfold in front of those flickering ponderosa groves was an aw-shucks celebration of patriarchal pride—one, by the way, that has plenty to do with our current environmental crises. If you were the right sex and the right color, you, too, could be out there riding into the sunset, though more often than not riding on a blinded horse.

By the time I was a teen, I was having serious doubts about whether I'd ever stack up to the boundless confidence of the swashbuckling Cartwrights. Given my life in a working-class Midwestern neighborhood, waking up every day to a patch of range less than the size of a tennis court, not sure who I was or what I was supposed to become, I doubted I'd ever muster such self-assurance. Curiously, that realization left me hankering even more for those shining ponderosa forests on the flanks of the Sierras. For reasons that still escape me, somehow I knew deep down those trees would be as happy to hang out with shy and shakey me as they'd been with Ben and Hoss and Little Joe.

~

For all the television prime time ponderosa was given, it showed up in plenty of other places too. In the middle of the twentieth century, the tree was making appearances in everything from French calendars to German oil paintings, the latter sometimes showing a single tree hanging like a dare from some unscalable sandstone tower. It appeared in no end of travel journals, too, from those of Lewis and Clark to John Muir to Enos Mills. And novels, including those by pulp fiction writers like Zane Grey—a man who first came to the ponderosa forests of Arizona from New York in 1909 on a harebrained expedition to rope mountain lions. Thirty years later, it was Louis L'Amour. Meanwhile, up in Montana, writer Norman Maclean used the tree to set the opening scene of his beautiful memoir, *A River Runs Through It:* "On the Big Blackfoot River, above the mouth of Belmont Creek, the banks are fringed by large ponderosa pines. In the slanting sun of late afternoon, the shadows of great branches reached from across the river, and the trees took the river in their arms. The shadows continued up the bank, until they included us."

Ernest Thompson Seton—by far the most influential children's nature author of the early twentieth century—not only spiked his pages with ponderosa but in 1930, at seventy, went to live among them in northern New Mexico. And in Marguerite Henry's beloved 1953 children's book *Brighty of the Grand Canyon*, when Brighty gets attacked by a mountain lion, it's a poultice made from the softened resin of a ponderosa pine that helps him heal.

Peter Fonda and Dennis Hopper rumbled on their choppers past groves of ponderosa in *Easy Rider*. And twenty years later, Louise throttled her turquoise '66 Thunderbird off a Utah cliff at Dead Horse Point, Thelma by her side; if they'd taken one quick glimpse to the south, their last look on the world would've included a blanket of ponderosa on the La Sal Mountains. And on it goes—Butch Cassidy, Indiana Jones, the Electric Horseman, Jeremiah Johnson. Even the amazingly popular contemporary TV series *Yellowstone* mostly showed the Duttons not near Yellowstone at all, but far to the west, in the Bitterroot Valley, playing out their swaggering lives against some of Montana's biggest ponderosa pines.

Given that long-standing urge we humans have to make connections with trees, especially trees of stature, it only makes sense that Western artists and storytellers gravitated to ponderosa. The ponderosa forest has an uncanny way of holding different moods—friendly to those who need a friend, enlivening to those who need inspiration, heartening for those who need assurance. Strong and beautiful and prolific, able to thrive alone as well as in vast forests, it's a tree that fits well into every kind of art and story we might wish to render.

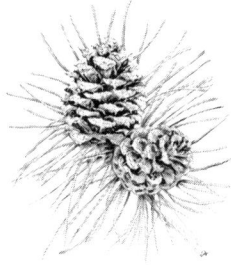

Chapter Six

I'm back on the highway, New Mexico State Route 4, heading south for the vast ponderosa forests around Jemez Springs. Along the way are sprawling scars from the 2011 Las Conchas Fire. At the time, it was the biggest wildfire in New Mexico history, eating up 153,000 acres; today, though, it ranks a distant fourth, well behind the 2022 Calf Canyon/Hermits Peak Fire northwest of the town of Las Vegas, which was more than twice that size. Notably, in 1900 the biggest tree-killing burns were rarely more than a thousand acres.

The Las Conchas Fire burned over 75 percent of beautiful Bandelier National Monument, in the process wiping out nearly all the park's ponderosa trees. In the first thirteen hours of the fire, some 44,000 acres went up in smoke, which pencils out to about one acre every second. To see the once-glorious Frijoles Canyon immediately after the fire was like looking at a sci-fi movie through night-vision goggles—an alien, monochrome world, the sooty black trunks of dead ponderosa rising out of dingy gray soil, drained here and there by sluggish, ash-colored creeks. In places the ash was knee-deep.

Today, thirteen years later, many of those places remain treeless. Not to say there haven't been some successful replanting efforts, as there've been across much of New Mexico. But the task is overwhelming. Of the more than five million acres that have burned in New Mexico in the early decades of this century, fully half of them will have little chance of turning into forest again without aggressive replanting. To complete that work would require a mind-boggling three hundred million seedlings, and millions more for the fires yet to come. The primary tree nursery in the region is the John T. Harrington Forestry Research Center northeast of Santa Fe, near the village of Mora, which has a capacity to grow three hundred thousand saplings a year. Meanwhile the newly established New Mexico Reforestation Center, which is slated for completion by 2028, will hopefully be able to grow five million. That's a lot. But keep in mind that the replacement trees necessary just to meet current needs would, with both of these programs running at full capacity, take more than fifty years to fill. Also, replanting ponderosa isn't a one-off proposition, where you just cruise the fire scars, shoehorn some seedlings into the ground, and call it good. In the Deschutes National Forest in the East Cascades of Oregon, only about 3 percent of viable ponderosa seed matured to two-year-old seedlings; and twenty-four months later, only those seedlings able to root under the cover of other vegetation were still alive.

Again, ponderosa tends to tweak itself to match the conditions of its home turf, which means the seed that does well in Bandelier may be less successful somewhere else. This is why most states in the West don't have just one seed bank but many, scattered across several regions. Back in the 1970s and '80s. the timber industry funded seed collection; by the end of the 1990s, that was rarely the case. In the wake of the Las Conchas Fire, an organization was formed to support seed-collecting efforts in the state—a hopeful effort involving the Nature Conservancy, the Wildlife Conservation Society, New Mexico State University, and the Santa Clara and Cochiti Pueblos. During one event in the fall of 2019, volunteers collected about 350,000 seeds.

In the places where the forest doesn't make it back, the plant community rising in its place is usually stitched with grasses and Gambel oak and locust shrubs. All of which have an easier time holding on in harsh conditions than a forest does. For one thing, grasses and shrubs have the very big advantage of being able to resprout from rootstalks. When a fire like the Las Conchas comes along and burns up the trees, the burn opens the land for shrubs and grasses to walk in on the ashes. In what's now emerging as a familiar pattern, a few years later another fire will come, this time taking the trees that survived the first burn along with any young saplings that may have sprouted. After every subsequent fire, the shrubs and grasses will rise from their roots, in time becoming the dominant ecosystem.

When I walk out to explore a slice of the burn scar west of Bandelier's visitor center, I find myself in one of these open grass and shrublands. What I notice first—and what's very different from my last visit here back in 2000—is the slap of a hot, hard-driving wind. In what for thousands of years was a landscape crowned by towering green pines, shielding all manner of creatures from various kinds of weather, is today a place where the winds run strong, full of dust and pummel. Two dozen years ago, I could stand at the edge of the woods and stare off into what were mostly waves of ponderosa forest in the surrounding uplands. But now—and probably for the first time in centuries, if not millennia—many of those same mountain flanks offer little clue of the trees that once made their homes there.

~

Before we launched our misguided efforts to suppress all wildfire, the kind of burn most common in the West was one that rolled through the forest with seven-to-ten-foot-high flames, typically producing temperatures of around 1200 degrees. Today, though, it's not uncommon to have wildfire with one-hundred- or even two-hundred-foot walls of flames burning at 2000 degrees—thundering up and over entire mountain ranges, even crossing major rivers, including, incredibly, the mile-wide Columbia. Fires so hot they can sterilize the soil, sometimes even destroying the underground fungal networks that trees are so dependent on. Fires fierce enough to vaporize plants, in the process coating everything with a silicone-like sheen that makes the ground hydrophobic, no longer able to absorb water, which can make much worse the problem of flooding and landslides.

The unprecedented intensity of today's fires often makes it hard for a burned forest to rebound. Indeed, in one biological survey, scientist Camille Stevens-Rumann was stunned to find that in well over a third of the ponderosa forests burned in the Rockies since 2000, not a single young tree was growing. In Colorado, more than a decade after the Hayman Fire roared through the ponderosa northwest of Colorado Springs, a fifty-thousand-acre swath near the center of the burn is still mostly without trees. The same is true in large swaths of Arizona and New Mexico and Southern California.

Besides the heavy fuel loads of dead wood our overeager fire suppression left behind, the practice also meant that saplings in the forests were no longer being thinned out by those regular modest burns. The forests became much more crowded. Research by dendrochronologists has found that before European settlement, ponderosa pine in central Arizona—an ecosystem not unlike this one near Bandelier—grew at a density of about fifty to a hundred trees per acre. Now that same acre may have five thousand trees. That, all by itself, has led to more intense fires—fires that climb into the crowns and kill even mature trees. Again, overenthusiastic fire suppression isn't the only reason ponderosa are dying in such vast numbers today. But it's a big one.

So then this question: Given the importance of wildfire to the health of a western forest, how did we become so thoroughly sold on the idea of stopping the woods from catching fire?

Our distaste for fire in the West goes back a long way. In 1879, John Wesley Powell—that illustrious, heroic one-armed explorer of the Grand Canyon—declared that the protection of forests across the arid regions of the West could be "reduced to one single problem: Can the forests be saved from fire?" Fires from lightning, of course. And as time went on, fires from railroad locomotives and loggers and random travelers. But what drew the most ire from policymakers were the fires intentionally set by Indigenous tribes—which by this point they'd been doing for thousands of years. John Muir called the cultural burning practices of the tribes the great scourge of the western forests, and then he called for armed soldiers to be brought in to stop them. Such indignation seems especially strange when you consider that most white settlers, from ranchers to farmers and even loggers, were themselves routinely burning the land.

Tribal burns were done in part to create better conditions for grazing animals, first wild and then domestic, as well as to support sun-loving food and medicinal plants, from huckleberry to oak trees to milkweed. Out on the prairies of the Great Plains, near the Black Hills, hunters sometimes used fires to guide the movements of bison. To Indigenous cultures, then, fire was a tool. Fire was medicine. Fire was a gift. Yet for many in the emerging science of forestry during the early twentieth century, the idea that Indigenous people with no scientific training—at least, no training recognizable to these European transplants—might have something worthwhile to share about taking care of the land was, well, frankly ridiculous.

The Annual Report of the Secretary of Agriculture in 1910 came down hard on a suggestion made by a small group of researchers daring to call for a few forests to be burned every year or two to remove ground debris. "As a matter of fact," the report sniffed, "such fires were enormously destructive. It is inconceivable that there should be seriously advocated a treatment of the forest that would inevitably result in the very rapid diminution of its density to the point where ultimately there would be no timber at all." The following year saw passage of the Weeks Act, which made "cultural uses of fire" officially illegal.

No one could accuse the government of not being tenacious. We kept our fists clenched against wildfire for almost seventy years, with the top brass playing whack-a-mole with any land manager who stuck his head up and argued for burning. Harold Weaver—forest supervisor of the Colville Indian Reservation and a nearly unrivaled expert on

ponderosa—wrote an article in 1943 for the *Journal of Forestry* coura-
geously arguing for the profound ecological value of fire. It's obvious,
said Weaver, that the present policy of attempting complete protection
of ponderosa pine stands from fire was causing big problems. "The
present stagnated stands of ponderosa have obviously resulted from
total fire exclusion."

He couldn't have been more right. Yet he, too, met a brick wall of
opposition. Forest Service scientist Arthur A. Brown responded to the
article, summing up the official attitude: "To serve society, the forester
must substitute harvesting by logging for nature's method of harvesting by pine beetles and fire. To do that, he must intervene in the old
natural cycle."

The war against wildfire got a shot of energy in 1902, with the
seven-hundred-thousand-acre, human-caused Yacolt Burn in western
Washington. Loggers working near the Columbia River Gorge managed to save themselves from what looked like certain death by running out of the ponderosa forest and diving into the waters of Trout
Lake. Nonetheless, thirty-eight people died, while nearly 150 families
lost their homes. At 238,000 acres in size, it would hold the record for
the state's biggest blaze for 112 years.

But it was eight years later, in 1910, when the anti-fire crusaders
were given their best chance to turn wildfire into public enemy number one. It was an event known simply as the Big Burn. Fires had
come early that year to the Northwest following a dry winter, with
April seeing small blazes already kicking off across a wide area. By
June the mountains along the Montana–Idaho divide were being hit
by frequent lightning strikes, many touching off fires pushed by hard
winds across bone-dry lands. Then still more burns sparked up—some
from lightning, others from loggers. More than a hundred blazes were
sparked by steam locomotives alone.

Suddenly there were hundreds of wildfires running across thousands of square miles of the inland Northwest. The Forest Service
was still young, just five years old, and eager to prove its worth to

lawmakers—many of whom thought the new agency was just more bloated government. Yet eager as the agency was, in the face of the Big Burn, it was soon overwhelmed. Rangers began scouring small towns near the fire zones, imploring every able-bodied man to grab an axe or shovel and head for the fire lines. Meanwhile, the higher-ups in the Forest Service grew so desperate they finally reached out to President Taft for reinforcements. He refused. They asked again. Still no. So they tried again. In the end Taft relented, ordering four thousand federal troops to join the suppression effort, including the legendary Buffalo Soldiers, made up of the Black infantry and calvary units created by the army shortly after the Civil War. After some harrowing, backbreaking efforts, by the third week in August, things were improving, as the biggest of the fires started to lay down. The worst was over. Thousands of men wiped the smoky char off their faces, grabbed their few belongings, and headed for home.

Then came August 20. The day of the blowup. A fast-moving dry cold front roared in on seventy-mile-an-hour winds, fanning the dwindling fires into flames more than two hundred feet high. The resulting heat created such strong updrafts that some men would later report seeing entire burning pine trees plucked out of the ground and sent twirling like death across the smoke-covered landscape. As one forester put it, the fire was "a veritable red demon from hell."

Three days later, on August 23, the air began to cool, then fill with thin curtains of rain and, at higher elevations, even snow. This time, the Big Burn really was winding to a close. In just thirty-six hours, three million acres had gone up in flames. Eighty-six people were dead. The town of Wallace, Idaho, was almost destroyed. And to the dismay of many newly minted foresters and forest rangers, enough trees were lost to fill the boxcars of a freight train 2,400 miles long.

The fledgling Forest Service wasted no time making its case. No one else, they argued—not miners, not railroad workers, not loggers—gave a damn about wildfires. The job would have to be handled by the US Forest Service. Seizing the moment, Forest Service Chief Henry

Graves came out swinging for the fences. With manpower and scientific prowess, he told reporters, his agency would eliminate fire from the landscape. More than a dozen years later, as several prominent conservationists, including Aldo Leopold, as well as the Forest Service's own fire control officer, proposed letting fires burn in the backcountry, the answer remained a resounding no. In fact, it was then that Forest Service Chief Ferdinand Silcox implemented the "10:00 am rule," setting a goal of extinguishing all new wildfires by ten o'clock the morning after they ignited. An ambition that in any era, in almost any place in the American West, is a fantasy of spectacular proportions.

~

And then there was the bear.

On a cool May morning in 1950, when telephones were still nowhere to be found in the Indigenous community of Taos Pueblo, the chief of the village climbed up the tall rock formation at the edge of the village and called out in the Tiwa language for all to hear. There was a fire to the south, he said, on the Lincoln National Forest. It was spreading fast, and twenty-five men were needed to help fight it. Those willing and able were told to gather in the village plaza.

It was only the second chance at firefighting for the newly formed Taos Snowballs—a group of Native men who, on first donning the wide-brimmed white metal hats that had been found for them in the back of a fire cache, glanced at one another and decided they looked rather like the snow-covered peaks of the Taos Mountains. Snowballs. Inexperienced as they were, they were eager, and they were tough.

The bus was loaded, the door pulled shut, and the men started rolling south, swaying along the twisting Rio Grande, brown and frothing with snowmelt, heading as fast as a school bus could travel toward the Los Tablos Fire. Many were wrapped in blankets against the cold, munching on tortillas filled with pinto beans, potatoes, and onions. That meal was followed a couple hours later by a stop at the popular

Lindy's Diner in the small city of Albuquerque, on Fifth and Route 66, where the crew was handed apples and bologna sandwiches.

It was early evening when the Snowballs finally rolled into base camp, and there given shovels and axes and Pulaskis, along with canteens of water. And still more bologna sandwiches. They climbed into army trucks and were driven to the drop-off point, where they hiked several miles up a mountain to the fire line, relieving a day crew made up of soldiers from Fort Bliss, Texas. They worked hard through the night, widening and expanding the fire lines, ever on the alert for hot spots being ignited by embers flying about on the spring winds.

By late the next morning, the Los Tablos Fire was mostly contained. So the Snowballs gathered their meager belongings and readied to board the bus back to Taos. And then suddenly, word came that another fire had kicked up. And this one—soon named the Capitan Fire—was making a fast march right for the base camp, eating up high-elevation forests of Douglas-fir, aspen, and white fir. The winds were sporadic, but now and then furious. Though at this point the Snowballs had only been off the fire line for a few hours, they loaded up their tools and headed back out.

Meanwhile, up on the mountain where the new fire was raging, a group of Fort Bliss soldiers under the direction of state game warden Speed Simmons was in trouble, trapped in a narrow canyon directly in the path of a fire front running right for them. Just as the inferno was about to overtake them, Simmons ordered the terrified men to lie face down in a nearby rockslide, burying their faces in the crevices of the rocks where, with luck, there might be enough cool air to keep their lungs from burning up. The flames were so close now; loud explosions were coming from the fire crowning in nearby Douglas-fir trees. Some thought hiding their faces in the rocks was a hopeless idea. They did it anyway. And incredibly, lying face down in those piles of rock, their noses poking hard into pockets of cooler air, left them with only minor burns and a lot of singed clothing. Visibly shaken, they were making their way down the mountain back to base camp when they met the Snowballs heading up.

Before the two groups parted ways, Speed Simmons told the Snowballs that his men had come across a small, terrified bear cub. He said a couple of the soldiers tried to catch it, but gave up after the little bear put his teeth and claws to work on them. Sure enough, when the Snowballs reached the slide where the soldiers had barely escaped death, in the fading light they spotted the cub whirling away to take refuge a few feet up a charred snag. Turning their headlamps on the little bear, they could see the hair on its ears and backside was singed, along with paws so badly burned they were already giving rise to a painful-looking crop of blisters. Puzzling it out, they decided the cub probably survived by taking refuge in that same rockslide, right along with Simmons and his crew, burning its paws only later as it made its way across the hot forest floor.

One of the Snowballs managed to grab the little bear, but again it put up a hearty fight. Once back on the ground, it trailed behind the crew as they made their way up the mountain, keeping with them for several hundred yards before finally giving up. Later on, someone did finally manage to capture the cub, though it's not exactly clear who—or how. Maybe it was a firefighter from the Mescalero Apache Red Hats, or from the Santo Domingo and Zia Pueblos. Either way, the cub was taken back to base camp, where he was christened Hotfoot Teddy. The Taos Pueblo firefighters, the Snowballs, would later say they felt a strong connection to that half-burned cub.

Hotfoot Teddy was soon being flown by small plane to Santa Fe, where a veterinarian treated his injuries. Recovery was slow, overseen by game warden field chief Ray Bell, along with his wife Ruth and their kids Don and Judy. The tiny five-pound bear would get no end of attention from the family, including being given an almost bottomless bowl of pablum cereal and honey. One newspaper reporter described how Teddy was "making the best of civilization." It proved a great local news piece. But then the story broke clean out of New

Mexico, spreading newspaper to newspaper across the country like, well, wildfire.

Six years before the Capitan Gap Fire, the Forest Service had commissioned renowned animal artist—and soon to be *Saturday Evening Post* cover artist—Albert Staehle to create a poster of a cartoon bear named Smokey, who was to help spread the word about wildfire prevention. The agency considered wildfire prevention especially critical at the time, given that millions of able-bodied Americans were off fighting in World War II, depleting the ranks of fire crews across the West. The best way to save the forests—forests described as essential to the war effort—was to keep fires from starting in the first place. Cartoon Smokey delivered his first catchphrase in 1944: "Care Will Prevent 9 Out of 10 Wildfires." It would be three years before he started speaking in more sizzling sound bites, reminding us in the end that "Only YOU can prevent forest fires."

Hotfoot Teddy seemed a golden opportunity for Smokey to come to life. As far as anyone knew, Teddy no longer had a mother—the assumption being that she'd either perished in the fire or wasn't able to return across the burning ground to retrieve her cub. Releasing him back to the wilds by himself, with no mother to teach him how to survive, was out of the question. So when his recovery was complete, the New Mexico Game Department made a gift of the now renamed Smokey Bear to Forest Service chief Lyle Watts, under the stipulation he be employed for conservation and wildfire prevention. In early July, when most little bears and their siblings were up in the mountains stumbling after their mothers, wrestling in the grass, and feasting on alumroot and cicada nymphs, the real Smokey Bear was high in the sky, making his way by airplane to the National Zoo in Washington, DC.

The real Smokey Bear lived to the ripe age of twenty-six, dying in 1976 of natural causes. It's impossible to overstate his impact. While at the National Zoo, he routinely received ten to fifteen thousand letters a week—so many, in fact, that the Postal Service gave Smokey his own zip code, which until then was something made available only to presidents. (It's 2025, by the way, and it's still in use today.) There were Smokey Bear dolls, and there were radio shows with Smokey talking to major celebrities, from Bing Crosby to Dinah Shore to Roy Rogers. He even got his own postage stamp. Senate resolutions were passed in his honor. License plates were issued. His death was front page news in the *Wall Street Journal*, and there were major obituaries in *The New York Times* and *Washington Post*. To this day, he remains the most recognizable public service icon in the world.

Yet maybe every hero, even bears—whether cartoon or flesh and fur—ends up carrying some baggage. The real Smokey came online four decades after the Forest Service declared war on wildfire. Yet for his whole life, people used him to help keep that war in full swing. Smokey unquestionably helped prevent lots of careless human-caused fires, especially given his popularity as millions of people began pouring into the woods for recreation following the end of the World War II. But he also helped entrench in the American psyche a strong resistance to the notion that any fire in the forest could serve a purpose, that it could be anything but destructive.

In the interest of full disclosure, I should mention that I actually played Smokey Bear for a brief moment back in the 1980s. One summer day, while we were living for a short time in Phoenix, my first and late wife, Jane, floated what seemed like a fun idea. She'd just landed a job teaching at a day-care center. Being a playful teacher, she thought it would be entertaining for the kids to have a visit from Smokey—which was to be me. Having both worked for the Forest Service, we knew that administrative offices often had Smokey Bear outfits they made available for outreach. Sure enough, the Tonto National Forest headquarters had just such a costume, and they were willing to lend it.

Early in the afternoon on the big day, I pulled on that thick, fuzzy wool suit and climbed into the van. A van with no air-conditioning. It was 117 degrees. Five miles of heavy traffic followed, punctuated by mild panic and waves of nausea. On reaching the school, I piled out the door to pant in the parking lot for a couple of minutes, managed to keep from throwing up, then slipped on the head of the costume and walked in the front door.

Smokey Bear, it turns out, is not universally beloved by tiny children. With the exception of two fearless little girls, every kid in the room screamed like the monsters in their closets had come to life. There was a lot of crying and a lot of hugging. Three boys ran off together and hid in the art closet. I was asked to leave.

~

Like most ecologists, forest scientist Craig Allen at the University of New Mexico is well aware that in any ecosystem change is a fundamental part of life. Still, when thinking about the ponderosa-rich Jemez Mountains of western New Mexico, he admits to feeling a troubling sense of loss over "the disorienting, rapid conversion of so much old conifer forest to non-forest—over the widespread death of so many familiar and favorite beautiful trees, from magnificent overstory titans to ancient, gnarled dwarves." Allen also mourns the additional losses he expects to be coming soon—"what my grandchildren likely will never be able to experience and might never even know is missing." Sadly, today the ancient Tewa village of Abiquiú, at one time rich in ponderosa, has lost thousands of them. Meanwhile the once aptly named Los Pinos watershed is woefully short of pinos.

After a night spent just outside Valles Caldera History Grove, my next stop is the very area Allen is talking about: the stunning Jemez Mountains, near the town of Jemez Springs. While settlers and sheepherders and ranchers and loggers and forest rangers have all left their tracks on this landscape, it is first and foremost the home of Indigenous people like those of Jemez Pueblo—residents here today, as they have been for more than a thousand years. And it's been a very good home, to be sure, with ponderosa and pinyon forest, flowing water, animals to hunt, and soil for farming, not to mention some of the best obsidian outcrops of the entire region, perfect for making scrapers and knives, spear points and arrowheads.

A thousand years ago, the larger landscape of northern New Mexico was in fact home to tens of thousands of people. Part of the longest-studied archaeological region in the country, there are yet today countless remains of ancient dwellings and spiritual gathering places, including magnificent cities of stone that reached three and four stories high, encompassing several thousand rooms. (A quick word here about names: While a Native person living in one of New Mexico's nineteen Pueblos is sometimes referred to as "a Pueblo," and their predecessors as "Ancestral Pueblos," this is not a name they call themselves. The term Pueblo first shows up in reports to the king of Spain, written by Spanish explorers.)

The Jemez people lived for much of the year in a valley along the Jemez River, tending corn and beans and squash, collecting piñon nuts and the fruits of prickly pear. During summers some climbed out of the canyon to live in smaller one- and two-room houses here in the cool of these ponderosa forests—gathering herbs and berries and medicinal plants, as well as hunting deer and rabbit and wild turkey. The physical remnants of their lives are scattered all through the forest, showing up as low lines of stones from the collapsed walls of their small dwellings. Walking along the ridgeline, I see them almost everywhere I look. They create the feeling of a people long gone and yet still here, the echoes of their lives wafting with the breeze through the ponderosa canopies.

It's in this region of the Southwest where the profound practical importance of the ponderosa to early cultures becomes crystal clear. Not just as a source of food and medicine, but as an essential part of the architecture of nearly every structure. Just as thousands of years earlier the ponderosa had given itself for fashioning the roofs of modest pit houses, so too would it later provide the wood for crossbeams, or *vigas*, used to build larger adobe dwellings and gathering spaces—including those mind-boggling multistory cities. Ponderosa was the solution to a very big problem—namely, how to create a roof system strong enough to hold the weight of several levels of floors.

The issue for some communities, though, was that the nearest ponderosa groves were forty or even fifty miles away in the surrounding mountains. This was the case at Chaco Canyon, well to the west of here, which contained great houses like Pueblo Bonito, with some six hundred rooms. From what we can tell, more than 70 percent of the ponderosa used in Chaco's construction came from far away—initially starting around 850 CE, from the Zuni Mountains, some fifty miles to the south, and later from the Chuska range, a similar distance to the west.

The Ancestral Puebloans did what they had to do, traveling in groups to the mountains to cut just the right-sized trees with stone axes. Then, with people positioned around the trunks of the harvested trees, each of which probably weighed between two and three hundred pounds, they carried them back those fifty miles. In Chaco Canyon alone, created between 850 and 1200 CE, some 240,000 trees were brought in from these distant ponderosa groves. Along with those trees there once came, also from the Chuska Mountains, a twenty-foot section of a single mature ponderosa, 250 years old, cut around 1100 CE and carried into the heart of the Chaco complex and placed in the center of Pueblo Bonito. Even if the tree was dead and well cured when it was cut, that twenty-foot section would likely have weighed well over a thousand pounds. Its purpose remains a mystery.

The ponderosa, then, gave people much of what they needed to live here. But the one thing these forests couldn't help the people with, outside of offering an occasional means of escape, was to spare them the brutal treatment meted out after contact with the Spanish, beginning around 1540 with Francisco Vázquez de Coronado.

Coronado, like so many who followed him, entered this world dragging a full load of bigotry and blinding obsession for gold, shoring up his obsessions with soldiers armed with everything from matchlock rifles to bronze cannons, as well as no fewer than six different kinds of bladed weapons.

The battles that reared up shortly after Coronado's arrival came to be known collectively as the Tiguex War—the first major conflict between Native people and foreigners in what later became the United States. Fighting erupted after multiple offenses by the Spaniards against the Tewa people: stealing their food and clothing and then kicking them out of their homes in the middle of winter; turning Spanish cattle out to devour stalks from the village cornfields, thereby removing a critical Tewa fuel for cooking and warming fires. And, most significantly, repeated sexual assault of Tewa women. Those offenses would continue off and on, both with Tewa and other Indigenous cultures, for the next 150 years.

One especially notable outrage came in 1596, after soldier Juan de Oñate led 129 soldiers, ten Franciscan priests, as well as large groups of women, children, servants, and slaves into New Mexico's Rio Grande Valley. At the time about forty thousand Indigenous people were living in the region, in well-functioning, sophisticated communities, many of which had by then been around for centuries. At Acoma Pueblo, southwest of where I'm standing here on Virgin Ridge, Oñate and his soldiers tried to steal the people's food stores, an act that would've doomed them to starvation in the coming winter. Fighting back hard, Pueblo warriors ended up killing eleven soldiers.

Whether the act was killing soldiers or merely taking their sheep, the Spanish military tended to have one category of response: big and violent. Oñate wasted no time striking back at Acoma Pueblo, massacring eight hundred men, women, and children. Across the years other officials, including the Spanish governor, hatched gruesome deterrents for those even suspected of being allied against them—like when they severed a foot from each of three Pueblo leaders, then sent them back to their villages as a warning to others.

By 1680, after short periods of calm when the Spanish were busy elsewhere, followed by still more horrific abuse, the Native people had had enough. Coordinating in secret among the nineteen Pueblo communities—all of which, by the way, are still here today—a plan was forged for a major uprising. From the final secret planning meeting, runners were sent out to each village carrying knotted cords; each day a leader of that village untied one of the knots, and when the last one was undone, it was time for war. As Zuni historian Jon Gahate told me of the Pueblo uprising, "It wasn't merely a fight for freedom. It was a battle for the very right to exist."

Remarkably, that uprising in 1680 drove the Spanish out of New Mexico. It was the first and only time when Native people in what would become the United States managed to expel an invading foreign nation. When the Spaniards returned twelve years later, they showed up helmets in hand, promising to be well-behaved, offering up what was essentially a power-sharing agreement. And for a while, things were better. But it didn't last. Here in Jemez, soldiers made a siege of Pueblo Mesa, killing some eighty Indigenous warriors. To avoid being captured, some of those warriors jumped off the towering cliffs on the flank of the mesa. According to a legend still told today, many were brought softly to the ground, as if being carefully lowered by invisible hands, said to be the grace of Saint Santiago. Following the battle the Spaniards rounded up several hundred old men and children and marched them off to Santa Fe, their backs bent and aching under the heavy bags of food stores stolen from their village.

~

Archaeologists have been working to unravel the mysteries of these cultures for almost 150 years. Sadly, for a long time those doing the unraveling included what might be best called scientific grave robbers: another band of outsiders looking for treasure. It was then a common practice among archaeologists to open graves and cart the contents off to far-away museums. Some also chose to share with the outside world sacred symbols, as well as ceremonies and stories they'd been explicitly told not to reveal. The Zia Pueblo sun symbol, for example, which today adorns the New Mexico state flag, was a sacred image on a pottery jar belonging to a Zia secret society, stolen by anthropologist James Stevenson in the late 1890s. Years later, it showed up on display in Santa Fe. When, in 1923, the state held a contest for a new state flag design, archaeologist Harry Mera and his wife submitted this sacred symbol. They won the contest, and two years later it was adopted for the state flag.

Besides all this theft and appropriation, the methods of early archaeology also had some enormous blind spots. Forty years ago Native archaeologist Rina Swentzell, from Santa Clara Pueblo, dared to point out the elephant in the room: Why were Anglo archaeologists in the Southwest so reluctant to use the knowledge and philosophy of the Pueblo people when interpreting a culture's past? Could Western scientists, who thought of themselves as paragons of objectivity, stop long enough to consider that their interpretations of what they were finding in ancient ruins might be deeply colored by their own worldviews?

The departure of the people from Chaco Canyon during the twelfth century, for instance, was claimed by some Western archaeologists to be the result of the people having exhausted their environment, including cutting down the ponderosa forests that supposedly grew nearby. To a Western mind, that's an especially reasonable interpretation given we come from a culture steeped in centuries of overuse and abuse of the environment. This isn't to say such overuse never happened in Indigenous cultures. But there were no such forests at Chaco, and why the people left is by no means a settled question. As Rena Swentzell points out, to the Indigenous people of this region, the world is in constant flux. Rather than settling in a single place, they "emulated the movement of the seasons, winds, clouds, and life cycles by moving frequently. They responded to the movement of floods, droughts, and social tensions." The process of "becoming" is for them an unending story, never complete.

Early archaeologists were also hobbled by one of the steadfast commandments of Enlightenment science, which is that there must be one absolute truth. Not so for the Native people here, or for that matter in many other places, who believe that instead of a single truth there are many, just as there are multiple levels of being.

Traditional Western archaeology liked to divide ancient villages into distinct parts: the dwelling area, the ceremonial site, the farming ground, the distant forest. For the Pueblo cultures, there is no separation whatsoever between village and people and landscape. Similarly, archaeologists assumed they were excavating parts of history, and those parts would allow them to forge an intellectual story of the past. The Indigenous people, on the other hand, say they experience history not as single events stuck in the distant past but as something alive in the present, here to offer clues for what action may be needed right now. Rena Swentzell summed it up well: "Philosophically, Southwest archaeologists and Anasazi [Pueblo] people are from distinct worlds."

I'm cautiously heartened by how one of those "distinct worlds"—the one made up of the dominant culture—may be inching toward some greater capacity to recognize the wisdom of others. No people could have established a civilization and thrived for more than a thousand years in a demanding landscape like this one without having an incredibly deep base of knowledge. With that in mind, ecologists have recently been working with elders from the region to better incorporate traditional cultural burning practices into modern wildfire management—something that could prove an important step in saving parts of the ponderosa forest.

Several times on this journey I've found myself wishing for a glimpse of those long-ago days before the great ponderosa groves of the West fell to the saw; before the days of European contact. My Tewa friend might bring me back to present by telling me how this world is in a constant state of transformation and re-creation. And that the work is about keeping continuity—about carrying forward the bonds of land and community that sustain and nourish us. Even in the middle of what might seem like catastrophic change.

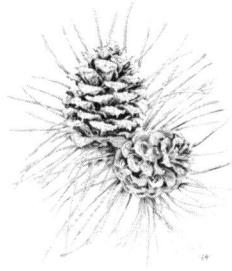

Chapter Seven

Over the past hundred years especially, no matter what culture an archaeologist or even climate scientist may come from, one tool useful to nearly all has been dendrochronology—the deciphering of the past through the stories held in the rings of trees. And as it happens, in much of the West the trees where such stories are most often revealed are ponderosa pine. Not just because ponderosa is so common. But also because compared to most other trees, the rings lain down by this one are especially clear and discernible.

The person credited with taking tree ring science into the mainstream, today known as the father of dendrochronology, is Andrew E. Douglass. Born in Vermont in 1867, Douglass was one of those people able to master nearly anything he turned his attention to. When he finally settled into dendrochronology, he made it unimaginably cool. Like fifty full pages in a single issue of *National Geographic* cool.

When Douglass started his career, he wasn't aiming for anything of the sort. Back then his attention was mostly on the heavens—much like his grandfather, an amateur astronomer who made the first detailed North American record of a meteor shower. Intrigued by such fondness for the heavens, Andrew enrolled in Harvard to study astronomy. And physics and geology too. An eager, brilliant student, on graduating he was hired by the wealthy, deeply eccentric astronomer Percival Lowell, who in 1894 sent young Douglass on a journey west to Flagstaff, Arizona—charged with finding there in that high, clean air a suitable place to build a telescope. One that could be used, among other things, to keep track of Lowell's growing obsession with the planet Mars. Settled in Flagstaff, Douglass would go on not only to engineer and build telescopes but also to come up with some nifty breakthrough methods for photographing the sky.

Though his boss's passions were wrapped up in Mars, and for a while Douglass would study the planet as directed, his own interest lay in sunspots—specifically, what effect they might be having on Earth's climate. Finding historic records of sunspot activity was the easy part. Generations of sky watchers had made careful records since at least the days of Galileo, finding that sunspots tended to strengthen and weaken on a cycle of roughly every eleven years. But again, Douglass wanted to know whether they were influencing our planet's climate. And climate records in northern Arizona were scant, reaching back only twenty or thirty years. With sunspot cycles happening only every eleven years, a twenty-year climate history was of little use.

In 1901, seven years after his arrival in northern Arizona, Douglass parted ways with his boss. Or to be more accurate, Lowell fired him. Mostly because Douglass disagreed—like almost every other astronomer did—with Lowell's idea that Mars showed irrefutable evidence of canals engineered and built by aliens. For Douglass, there seems to have been no love lost. If anything, it allowed him to focus more on the climate–sunspot connection.

In the area near Flagstaff where he'd chosen to set up the observatory, aggressive logging operations had been going on since the 1880s, when massive numbers of ponderosa trees were felled to make railroad ties. A mile of railroad track took about twenty-five hundred ties per mile, and each of them had to be replaced every five to seven years. By 1900, over fifteen million acres of forest across the West, much of it ponderosa, was being leveled for railroad ties alone. The good news in all this for Douglass was that he could hardly walk a hundred yards in any direction without running into fresh stumps, many of them from three- and four-hundred-year-old trees. Taking note of how changes in climate showed up as changes to the width of the rings—as Leonardo da Vinci had done—Douglass figured he might be able to correlate those growth patterns with sunspot records.

As it happened, his results were inconclusive. Even today the results are a bit murky; at the moment many scientists think that a strong sunspot year probably accounts for just 2 or 3 percent of climate warming. That said, by using new methods of analyzing radiocarbon locked over thousands of years in tree-ring samples, scientists from the University of Arizona and ETH University in Zurich were recently able to create an extremely detailed record of solar activity during the first millennium BCE. Not only is this allowing a better understanding of the sun's influence in the past, but it may also help us predict future magnetic storms, which can wreak havoc with everything from power grids to global positioning systems.

Douglass began making records of the ring patterns he found in different trees, doing so in much the same way dendrochronologists do today. Say you want to begin with a standing live tree. Using a small, tubular metal tool called an increment borer, you drill out a sample of the tree (this doesn't harm the tree), which comes out as a cylinder about the size and shape of a drinking straw. Now draw a horizontal line on a piece of graph paper and, working left to right, make small vertical hash marks along the printed lines of the paper, with the height of each mark corresponding to the relative size of a particular ring. The narrower the ring, the taller the mark. Generally wetter years produce wider rings, while the rings in drier years are narrower. Now you have a growth map for a single living tree; to know when it first started growing, you can simply count the rings backward from the present day.

Say the next tree you choose is an old ponderosa long dead, maybe even an old stump. While a core sample will tell you how old the tree was when it died, you'll have no idea *when* it died—or, conversely, when it started growing. Once again, on another piece of graph paper make the same kind of horizontal map you made before, with the height of the hash marks corresponding to the varying widths of the rings. Now, hold one graph directly above the other one, moving it back and forth to see if you can find anywhere where the growth patterns you recorded for the two trees exactly match. In this case you might find that the right side of the graph you made for the old, dead tree—the right side of the paper representing the later years of its growth—matches perfectly with the early growing years of the living tree. Tape the two sheets of paper together at the match points, and presto, you now have a historical record spanning the lives of both trees. You know both the year that the old tree died as well as when it first started growing.

If you don't have any ring patterns that match, that means there's a gap between the life histories of the two trees. If you're really committed to creating a long, unbroken historical record—maybe, for example, to study changes in climate across many centuries—then you're going to have to keep examining trees until you find one that fills that gap. And that can take an almost ridiculous amount of looking.

This is the work Douglass was on about. Adding data from other tree species, most notably the sequoia, he eventually was able to look back across some three thousand years. The catch with the sequoia, though, is that while it could tell him about climate conditions on the west slope of the Sierras, it was of no use figuring out climate puzzles elsewhere, including northern Arizona, since the two regions have very different climate histories. Master maps would have to be built for every region of the West.

In 1914 an archaeologist working at Chaco Canyon, Neil Judd, happened to hear about Douglass's work and approached him for help. Would he be able to put a date on one of the ponderosa roof beams they'd found at the site? At the time, the master record that Douglass had created for the region went back only about five hundred years, at which point it hit one of those gaps mentioned earlier. The roof beam he'd been given fell into that gap. This meant that while he was able to tell the archaeologist that the beam was over five hundred years old, he couldn't say how by how much.

As far as the archaeologist was concerned, though, just knowing that the beams were more than five hundred years old was a triumph. Until that point no one had any idea about the age of Chaco Canyon or, for that matter, any other of the region's ancient settlements. Sensing the value of what Douglass was doing, more and more archeologists started sending him wood samples. When a portion of a ponderosa roof beam came in from fifty miles south of Chaco, at what was called Aztec Ruins, he was able to find matching sections between the Chaco and Aztec samples; once he had that match, he could at least see that Chaco was older—that it was constructed before Aztec. And that bit of information, too, left the archaeologists practically giddy.

Douglass, though, was all about filling the gap. If he could just find that missing piece he could construct an unbroken master record—a climate record, if you will—stretching back more than a millennium. He was always on the lookout, roaming the woods of Arizona collecting samples, also combing places like Mesa Verde and Aztec Ruins. By now on faculty at the University of Arizona, he was awarded a National Geographic Society grant to help him keep up the search.

Ten years after Neil Judd asked for his help on that roof beam from Chaco Canyon, fate stepped in. Having heard of some old beams at an excavation site near Show Low, Arizona, Douglass—who was teaching classes in Tucson—climbed into a buckboard wagon and headed off on a long, bumpy, dusty ride. Barely had he climbed down from the wagon when one of the archaeologists handed him a piece of charred timber. As had happened countless times before, Douglass disappeared inside a makeshift shed set up as a primitive lab and workspace. There, under the light of a kerosene lamp, he plotted the growth rings of the sample, then held that graph up against what was by then a well-worn master record.

Two hours later, he walked out of that shed into the cool evening air and called together the excavation team. Looking by turns at each of them, he broke into a smile. This was it—this charred piece of wood was the missing link. The gap was filled. I can tell you, he informed them, that Chaco Canyon was built about 1130 and Aztec about thirty years later, in 1160. One by one he went down the list of the most studiously researched archaeology sites in the Southwest, dealing the dates of their construction to a slack-jawed group of scientists who could barely believe their ears, a group mindful of the fact that they were in the midst of one of the greatest moments, if not *the* greatest moment, in the history of American archaeology. That fifty-page article in *National Geographic* came out in 1929, titled "Secrets of the Southwest Solved by Talkative Tree Rings." Later, using ponderosa plucked from lava flows at El Malpais National Monument west of Albuquerque, scientists pieced together a continuous record extending back two thousand years.

While the value of such a record was evident to anyone wanting to date archaeological ruins, and also to anyone wanting to track climate change over the centuries, these ponderosa held other stories too. Those who really know how to look can find tracks left by big winds or lightning strikes, by insect infestations and fungal infections. There may even be traces of the bear that stopped by fifty years ago to scratch the trunk with her claws. Or fifty years before that, when some lovestruck young man carved his sweetheart's initials. Or even when some long-ago hunter missed his mark, his arrow gashing the tree.

Tree rings also hold records of the industrial age, which show up as traces of residue from coughing smokestacks. They carry traces of radionuclides from atomic testing, and no matter where in the world they happen to grow, from the bombs detonated over Hiroshima and Nagasaki. They hold the records of those rare but extremely powerful solar storms that infrequently hit the planet, known as Miyake events. And notably, when carbon 14 radiocarbon dating came along—a tool developed in atomic weapons research, used to date organic matter back about fifty thousand years—the technique was calibrated using tree-ring records.

The tree stories most fascinating to some scientists today, however, happen to be ones very relevant to the fate of ponderosa. And these are the tales of fire. Arguably, today no one is better equipped to tell those stories than the man standing with me here in the Jemez Mountains, side by side on the sheer edge of Virgin Ridge. His name is Tom Swetnam, and he's long been one of the country's most renowned dendrochronologists.

Swetnam is a jaunty seventy-year-old bear of a man with the quintessential look of a Western scientist. His round face sports a trim gray beard under a wide-brimmed, straw-colored hat. On hearing almost any of my questions about the forest, his eyes light up behind his silver, wire-rimmed glasses in a way that lets you know you're in for something good. While Tom's knowledge of the ponderosa forests—not just here but across the Southwest—is frankly astonishing, it's his enthusiasm that holds me. After a couple hours, I find myself musing about whether I should've become a scientist like Tom, puzzling out human and natural history from stories held in the trees.

We're nearly a thousand feet above the Jemez River. From here the whole wide world, as far as you can see, is stitched through with ponderosa. Some trees hang by their toes off the ridgetops, leaning out into thin air. Many others tell of wildfire—from three-foot scars on the bark of the tree right next to me to loose huddles of burned and blackened ghost trees miles away.

Peering over the edge of the ridge we can see Tom's house, some seven hundred feet below, nestled on the west slope of the very canyon where he spent his boyhood and now, in his post-university years, is back again. On the east side of his home is a spacious deck tucked into the ponderosa forest; on seeing it, I immediately imagine a perfect morning moment: stepping out there among the trees with a cup of coffee to catch the rising sun.

Despite having retired from the University of Arizona, Tom still has his fingers in too many projects to count. Behind the house is a small shed outfitted with a microscope, records of forest histories mapped out on graph paper, hundreds of thin horizontal slices from fallen trees, and what seems like thousands of thin, tubular tree core samples—some from last week, some from thirty years ago—carefully stored away in paper drinking straws.

Dendrochronologists like Tom figured out long ago how to use fire scars on tree rings to determine how often wildfire was running through the woods. Over time, it became clear that most of these burns fell into what today are known as "stand maintenance fires"—modest burns that have been part of healthy forests for tens of thousands of years. Ponderosa rings have taught us that those healthy fires came through the forest roughly every five to seven years, and also that around here, they were driven by a distinct wet-dry cycle. The pattern was to have two or three wet years, which led to grass and other plants forming thick weaves on the forest floor. That was commonly followed by two or three dry years, during which the vegetation dried out, making it easier for wildfire to spread through the forest. (This is one of several reasons why overgrazing has been so damaging to the forests: Devour all the ground plants, and a useful, modest fire loses its ability to spread.)

Thirty years ago, a scientific team that included Tom Swetnam released results from an ambitious study of fire scars in the ponderosa forests of New Mexico and Arizona. What they discovered is that way back when there were fewer humans in this region, modest fires burned through the forests every five to fifteen years. Then around 1300—an era that saw a big increase in human population, as the Jemez and other cultures began establishing larger communities—the burn frequency changed to just two or three years. For the most part those burns were smaller and more targeted, suggesting that like other Indigenous cultures across the continent, people here were intentionally bringing fire to the land, using it to encourage the healthy growth that favored food and medicinal plants, as well as to keep the grass lush for grazing for elk and deer. Then in the 1690s, the Jemez people were forcibly removed from their villages in the ponderosa forest by the Spanish, made to live lower down in the harsher desert steppe country. When they left the forest, the frequency rate of fires there went back to every five to fifteen years, which is where it remained until the early decades of the twentieth century, when the big national campaign to stop wildfire began. At that point, those long-running burn patterns disappeared, and the forests started their long decline.

~

The next morning brings that bright, hopeful New Mexico light pouring into Tom's kitchen like a prelude to the second coming. He's over at the stove being a line cook, his big hands shuffling skillets of bacon and eggs. Noting that he's very good at talking and cooking at the same time—something I can't begin to manage—I ask about some ongoing conversations going on in science circles about a small group of ecologists pushing back against the by-now common use of fire as a forest management tool, both to thin the forest as well as to burn up accumulated debris. Some of these ecologists—mainstream scientists often call them contrarians—would like to stop both prescribed burning and mechanical thinning altogether. They dismiss the effects of heavy fuel loads and high tree densities, claiming that the big, destructive fires we're seeing today are entirely the result of climate change. Thinning and prescribed burning, their thinking goes, make for yet another round of humans messing about in nature without really knowing what they're doing. The prudent thing would be to back off and let nature do what nature does.

I get where they're coming from. Long is the list of times when human beings around the globe have intervened in nature only to muck things up. We removed wolves and other predators from Yellowstone, inadvertently allowing elk populations to explode—not only rendering them more vulnerable to herd diseases but also weakening a key food source by degrading the range. Back in the Midwest we introduced Asian carp to control aquatic weeds, only to see the populations of that fish explode to the point that today they threaten native species in thousands of miles of rivers, as well as the Great Lakes. We built levees in Louisiana to control floods, in the process stopping the deposit of sediments carried by the rivers; today, every few hours a precious slice of wetlands the size of a football field disappears into open water. We imported fire ants from South America to control boll weevils and aphids; they quickly knocked entire ecosystems out of balance, proving to be one of the worst invasive species in the world. In 1956, Rachel Carson called attention to a spectacular boondoggle hosted by Idaho and Montana, which had undertaken the widespread spraying of DDT to improve crop yields. Hoping to keep crop- and vegetable-threatening critters like spider mites in check, they ended up instead killing off no end of important insects, including the ladybugs that ate the spider mites. Which led to, as Carson described it, "the most extensive and spectacular infestation of spider mites in history."

Tom, too, has some measure of sympathy for the contrarians. "Land management efforts have stepped in it over and over again," he says. "To put it kindly, it's a blemished record." He understands the temptation to buy the argument that because we screwed things up in the past, we're destined to do it again. And there are legitimate downsides to interventions like forest thinning: The roads that get put in not only cause erosion but are also pathways for bringing in the kinds of invasive plants that can wreak havoc with biodiversity.

"Still," he says, "we're in the face of climate change. This enormous stressor of our own making. We need to roll up our sleeves and keep trying. To my mind it's the absolute worst time for us to step back."

"Can we make a difference?" I ask.

"Well, I don't think we can save all of it. Maybe not even most of it. But we can save some of it." Keep interventions to a minimum, he says, and keep in mind that when it comes to ecology, we're not as smart as we think we are. "Engage as carefully as possible."

Is there still some risk we'll unleash unintended consequences? "Yes."

Despite the contrarian view, there's overwhelming evidence that unnatural fuel loads, along with the bunching of timber caused by fire suppression, are an enormous part of why we have such intensely destructive wildfires today. Granted, there should absolutely be places, and probably lots of them, where we don't do anything at all. But it seems a stretch to ask nature to heal after we've hung such a heavy load around her neck. Taking action in the woods, if done carefully, can be a lifeline for the forest. Prescribed burning especially—in many places now undergirded by the wisdom of Indigenous cultural burning techniques—has already led to forests rebounding where they otherwise wouldn't have.

"It's curious to me when I see a conservationist pushing to keep hands off the forests, no matter what," Tom says. "But at the same time, they're pushing to move native Masai off their lands in Tanzania to protect wildlife. Or in the name of saving sea lions, slapping injunctions on Northwest fishing tribes for exercising their legal fishing rights."

In the early days of modern science, the days of Descartes and Bacon, it was common to think of humans as good, nature as bad; or at best, nature as deeply flawed, clearly in need of our fixing. Now some of us have the opposite opinion: Humans are bad; nature is good. Either way we end up cleaved from the world, unable to reconcile the fact that we're nature too.

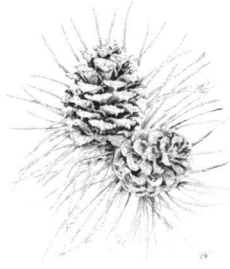

Chapter Eight

By the time I reach Alma, New Mexico, a faint speck of a town in west-central New Mexico at the edge of the Mogollon Mountains, the day has gone to swelter. The dirt roads are thick with dust, and even when I drive barely faster than a walk, giant belches of it rise up from the wheels and coat the branches of young ponderosa. I quietly apologize to the trees, knowing that some of the pores on their needles are getting clogged, which will in turn mess a bit with their ability to take in carbon dioxide.

Alma is often labeled as a ghost town, which hardly seems fair given that between forty and fifty people are living there. It is, though, clearly on the quiet side. Which is a far cry from what it was in the later decades of the nineteenth century. In 1880 Apache Chief Victorio traveled across these lands, at one point helping some eighty warriors and their families escape the brutal conditions of the San Carlos Reservation. Victorio was a superb military tactician, and with his tribe's homelands under increasing threat, he often found himself putting those skills into play. On a return journey back to San Carlos, hoping to liberate still other members of his own family, he ended up in a brawl known forever after as the Alma massacre. The conflict left forty-one miners, settlers, and soldiers dead, with no, or limited, casualties among the Apaches. Yet six months after that fight, Victorio was dead, killed by ambush in Chihuahua, Mexico. It's worth noting that his chief lieutenant at the time of his death was his brilliant, fearless sister, Lorenz, who some later came to call the Apache Joan of Arc. She narrowly escaped her brother's fate that day, having stayed back at camp to help a woman in labor.

Alma is also where Butch Cassidy and the Sundance Kid worked as cowboys for the WS Ranch. The ranch foreman who hired them fast became their biggest fan, having seen his cattle-rustling problems mysteriously drop away to almost nothing once the pair settled in. Here in Alma, too, was Billy the Kid's stepfather, William Henry Harrison Antrim—a prospector who, having shipped Billy and his brother off to live with guardians following their mother's death, roamed these drainages like so many others were doing, in a breakneck search for silver and gold.

On a far more peaceful note, thirty miles to the southeast of Alma as the hawk flies, on land awash in ponderosa, is the first protected wilderness in the United States—quite likely the first such preserve in the world: the magnificent Gila Wilderness. Created in 1924 from 750,000 acres of the Gila National Forest, this was the first step on a journey that would lead to America today having more than 175,000 square miles of protected wild country. That the Gila happened at all was in large part thanks to the impassioned efforts of forester and ecologist Aldo Leopold. Leopold wasn't just a brilliant thinker but also a scientist with an admirable willingness to admit his mistakes and learn from them. This was never clearer than when, after killing one of the region's last wolves in the ponderosa forests of Arizona, he later came to understand the critical role of predators in healthy ecosystems and denounced his own actions. In 1944, thirty-five years after that shooting, he became the first major conservation scientist in the country to call for wolves to be reintroduced into Yellowstone.

"Acts of creation are ordinarily reserved for gods and poets," Leopold suggested, "but humbler folk may circumvent this restriction if they know how. To plant a pine, for example, one need be neither god nor poet; one need only own a shovel. By virtue of this curious loophole in the rules, any clodhopper may say: Let there be a tree and there will be one."

A Midwesterner for much of his life, Leopold counted the lands of the Gila among his favorite places on Earth. It was thanks to this love affair that, as people worked through the twentieth century to safeguard wild places, a key image many held in their heads came from photos of these beautiful Southwestern uplands, wrapped to the gills in ponderosa.

A few miles outside Alma, I find a place to park in the woods for the evening. The heat lingers well into the night, leaving me lying in bed on top of the sheets with the windows open. Which is why, sometime around midnight, I'm able to easily hear the hoots and tootles of a great horned owl. She seems no more than ten yards away, all two feet of her, patiently watching and listening for others who go about in the night—able to hear the footsteps of a mouse walking on dry pine needles from a hundred feet away. She and her kin seem like perfect companions for the ponderosa forest, if only for how remarkably successful they've both been in finding homes across a staggering range of habitats. Indeed, hers is even more sprawling than that of the tree, stretching from West Coast to East Coast, South America to Canada.

Like many people, I've spent my whole life with echoes of the notion that owls are wise. I particularly like the story from the ancient Greeks, who situated the owl next to Athena, the goddess of wisdom—placing the bird on her blind side, thereby allowing Athena to perceive the whole truth of things. What if we mere mortals could forge that kind of fantastical alliance with a being that would help us see more clearly? The possibilities are endless. Maybe we wouldn't have jumped so fast to beat down every western wildfire. Maybe we would've steered clear of thinking we could snap our fingers and make nature bow to our will.

It's a little after dawn when I head off-trail from my campsite into the ponderosa forest. The woods are busy this morning, as if all the creatures are out and about getting stuff done before the heat turns up. Fifty yards in is a pair of Abert's squirrels, beautiful with their dark gray backs flushed with badges of russet, white chests, and ivory tails. Also called tassel-eared squirrels, for their long, pointed gray ears tipped with fine hairs, they look something like the offspring of a squirrel who fell in love with a rabbit. While a lot of creatures come and go from the ponderosa forest, making a living from several different environments, the tassel-eared squirrel spends its time nowhere else. Everything it needs and wants is in this forest, close at hand. The nest it builds sits high in a ponderosa—a small bowl of sticks, interlaced with needles for insulation. Unlike most squirrels, the tassel-eared doesn't really squirrel away much of anything. It's happy to eat whatever the ponderosa is serving up, from buds to cones to sap to nuts, even the inner bark.

The pair in front of me looks to be having the equivalent of a Sunday brunch, sitting on the ground nibbling a couple of boletus mushrooms at the base of an old fire scar. Like everything in nature, this meal isn't a one-way street. It's an exchange. The squirrels get a food source not only chock-full of protein and vitamins, but one that likely also gives a boost to their immune systems. The squirrel, in return, runs around helping propagate more mushrooms by casting spores in the forest every time it poops. Some of those spores go on to become critical parts of the mycorrhizal network, which the ponderosa absolutely depends on for accessing critical nitrogen and phosphorus, even helping it survive drought. There's a good reason why, when you hold the root tip of a ponderosa to your nose, what you smell is mushrooms.

Of all us creatures who love to eat mushrooms, Abert's squirrels are in a class by themselves. They even chow down on deadly poisonous species like the death cap without missing a beat, going happily on their way, while a mere ounce of it would in no time send any of us to an early grave. Exactly how they manage such a feat is still unknown. Some biologists suspect the squirrels evolved some kind of special mechanism to neutralize the amatoxins; others speculate they may have a special digestive process that breaks down the toxins before they can cause harm.

One reason the squirrels rarely pass a mushroom without nibbling on it is that it might not be there when they turn around. The mice and wood rats and chipmunks in these woods love them, too, as well as deer, elk, grasshoppers, worms, wasps—even birds. The bacteria in the scat of all those creatures gets mixed with the germinating spores, which makes for a beautifully nutrient-rich medium that helps push all kinds of plants from seed to seedling. The mushrooms growing here help feed the ponderosa, to be sure. But they also feed the grasses and shrubs that feed the deer and elk, which in turn feed mountain lion, whose kills in turn feed dozens of other creatures, as well as give nutrients back to those grasses and shrubs. This is a potent environment. Of the 450 species of mammals, birds, reptiles, and amphibians in this region, more than half have tied their lives in some way to the ponderosa forest.

~

Back on the road again, heading west. By early afternoon I've made my way onto the Mogollon Rim—a swooning, two-hundred-mile-long escarpment of limestone and sandstone cliffs at the southern edge of the 130,000-square-mile Colorado Plateau—which means I'm finally fully wrapped in the largest ponderosa pine forest in the world. Starting on the far side of the New Mexico border, the trees run west for almost two hundred miles, through Flagstaff all the way to the Grand Canyon, in a band swelling from twenty-five to forty miles wide. Here a person could ramble on foot at a decent pace for a full month before running out of woods. And if by then you still hadn't had enough, you could drop into the Grand Canyon, pausing in a hundred side canyons and ravines to admire lone giant trees clinging to the rock. On crossing the Colorado River and climbing back out again, panting up over a vertical mile to the North Rim, you'd end up on the Kaibab Plateau,

which holds one of the biggest remaining stands of old-growth ponderosa in the country. I have to say, just thinking about all this leaves
me happy as a dog with two tails.

Around four o'clock I reach the strikingly beautiful Rim Drive in
the Coconino National Forest, parking the van near the edge of the
cliffs in a loose toss of pines. From this eight-thousand-foot perch,
the world does a dizzy freefall, tumbling ass over teakettle through
the ponderosa all the way to the outer edges of honest-to-god desert.
In a state with no shortage of thrilling scenery, including the utter
shock therapy that awaits anyone who stands on the rim of the Grand
Canyon, the Mogollon Rim still qualifies as eye-popping. It presents
just the sort of scenes you might imagine in a good daydream about
the West, even if you'd never been here before, building your notions
about it from a loose toss of movies and calendar photos and clips from
public television nature shows. Even Zane Grey came here to conjure
up his fictional cowboys, writing them onto the page while sitting in a
hunting cabin he owned at the foot of the rim, on Tonto Creek. This is
deep West, iconic West, a stony land of steep cliffs and tight canyons,
where over the millennia millions of ponderosa have massaged the
rocks to life.

And yet almost none of the forest is old growth like the History
Grove, or the mature stands on the Kaibab Plateau, or even those of
the experimental forests around Flagstaff. What stretches out in front
of me, while gorgeous, is vastly different than what was here two hundred years ago. Wholesale logging, decades of hard grazing by cattle
and sheep, and of course fire suppression have led to crowded forests,
with a smaller quantity and diversity of plants in the understory. Fewer
plants on the ground means fewer small animals, which in turn means
fewer birds of prey, including goshawks and great horned and Mexican
spotted owls. What's more, before European settlement these forests
were broken in places by generous tosses of grassy meadows, bedecked
in late spring and summer with blankets of wildflowers; today, barring

a recent burn or insect invasion, the trees run mostly uninterrupted. Given the old maxim that you can't pull on one strand in the web of life without shaking all the others, it's safe to say that the world's biggest ponderosa forest has been seriously shaken.

These dryland pine forests of the Mogollon country seem like a good place to consider the second big reason ponderosa are in decline across the Southwest. That reason is climate change. Put the two together—fire suppression and a warming, drying climate—and you have a perfect storm that will continue to unravel this and other western landscapes for decades, if not centuries, to come.

Scientists were already predicting thirty years ago that the American Southwest would be hit especially hard by climate change: less rainfall, dangerously high temperatures, wildfires burning bigger and hotter. Over the last century the region has warmed fast—at least 2 degrees Fahrenheit, with even bigger increases in many of the places ponderosa grow. Some years back, UCLA bioclimatologist Park Williams made a deeply sobering comment, warning that "the majority of forests in the Southwest probably cannot survive in the temperatures projected." And that includes the ponderosa forests in front of me right now, sprawling south through Arizona from the feet of the Mogollon Rim.

On one hand, warmer temperatures diminish the snowpack, and less snowpack means less water for everything, including all the people living at lower elevations. In the twenty-one largest water reservoirs in the West, snowmelt makes up 67 percent of the storage. Climate change has also led to earlier springs, which dry out the ground and the dead wood and grasses of the forest, setting the stage for bigger wildfires. Flooding and landslides, too, have increased under climate change, as bigger rains from a warmer atmosphere fall on the bare soils of fire scars. Following one massive fire twenty years ago in the ponderosa forests of north-central Arizona, subsequent flooding scoured away soils that had taken more than eight thousand years to create.

As a boy, living downwind from the steel mills and oil refineries of northwestern Indiana, every now and then I'd hear adults complaining about the nasty odors wafting across the region. Most shrugged it off, though, with a few even calling it "the smell of progress." But as for the possibility that those belching smokestacks might be changing the climate, that notion wasn't on the radar of even most scientists. Those rare researchers who did express worry weren't just ignored but often scorned by their colleagues. In 1976 climate modeler Stephen Schneider, working at the NASA Goddard Institute for Space Studies, released a clear warning about climate change in his book *The Genesis Strategy*; many of his fellow scientists derided him for being an activist, something hugely frowned upon. Helmut Landsberg, the former director of the National Weather Bureau's Office of Climatology, ridiculed Schneider as undisciplined, painting his warnings as the work of an unserious scientist. Only a very few others—Carl Sagan was one of them—publicly supported Schneider's conclusions.

Even before these warnings, in the late 1960s and early '70s a shifting climate was likely already bringing changes, including weakening crop production in West Africa, China, India, Russia, and the United States. President Johnson had been advised way back in 1965 that unless something was done, carbon dioxide levels in the atmosphere could, by the year 2000, swell to levels 25 percent higher than they were before industrialization. (In fact, they would end up being 31 percent higher.) Johnson was concerned enough to write a letter to Congress about the matter, warning them that the atmosphere was being altered on a global scale "through a steady increase in carbon dioxide from the burning of fossil fuels." But it really wasn't until the 1990s that more scientists verified the data, and then started raising the red flag in earnest. And that, in turn, led more of the public to start paying attention. Which led the fossil fuel industry to put in motion the most expensive ad campaign in American history, denouncing the whole idea. And the clock kept ticking.

For those paying attention, it fast became evident that few places on Earth were likely to escape the effects of massive amounts of carbon dioxide being dumped into the atmosphere. As for trees in the American West, over the past thirty years, climate change is thought to have been a significant factor in bringing down about seven million acres of aspen forest, 85 percent of New Mexico's pinyon pines, most of the famous Yellowstone whitebark pines—the nuts of which are an essential food for grizzly bears—and thousands of ancient bristlecone pines in the Sierras, along with about one out of every five giant sequoia. As well as literally hundreds of millions of ponderosa pine. When hardy-as-hell trees like pinyon and ponderosa and even juniper start falling, the disaster is very much underway. All three of these species have thrived for thousands of years in places where being a tree is anything but easy: places of drought and heat and western pine beetles and ips invasions, of blister rust and root rot, needle scale and pitch moths and mistletoe. Yet all have been recently dying at a frightening pace—often in the heart of those wild places Americans fought hard for decades to preserve. Even with all this life burning up and dying of thirst, some people—including many at the highest levels of government—remain unwilling to even acknowledge the problem, let alone act. If hubris were combustible, they would burst into flames.

Climate change skeptics like to claim that humans have little to do with any of this, often pointing out that conditions on the planet have always fluctuated. As proof, many point to an era fifty-five million years ago, when rising atmospheric carbon dioxide ushered in 170,000 years of dramatic warming. Over that period many species went extinct, especially those living deep in the oceans. But what skeptics fail to acknowledge is that today's human-caused climate change, with the eventual high temperatures still unknown, is happening ten times faster. What's more, the changes we're seeing right now have human fingerprints all over them. Looking at the ratios of different isotopes in atmospheric carbon dioxide molecules, we can in fact sort out how much of that carbon dioxide comes from burning fossil fuels, as opposed to what was released by plants—or even what gets spewed out by volcanoes. Equally telling, the only way a computer model today can be made to replicate our current conditions is if the model takes into account the fossil fuels burned by humans. Leave that factor out, and the model will suggest a world far less ravaged than the one we see today.

∼

As it happened, two wildfire events in the ponderosa forests along the Mogollon Rim were part of what led firefighters and ecologists toward the understanding that not only was climate change real, but it was turning wildfire into something more catastrophic than it used to be. The first of those events, in June of 1990, was the Dude Fire. On the day the burn started, temperatures in the nearby town of Payson soared to 106 degrees, tying what was then the record for the hottest day in history. The Dude Fire was incredibly aggressive, consuming more than five dozen homes. It also trapped eleven firefighters from the Perryville prison under the Mogollon Rim; six inmates died, along with a prison supervisor, and five others were injured. Within a year of that tragedy, there emerged a new safety slogan in wildland firefighter training—one that still today gets drilled into every wildland firefighter's brain: *Look up, look down, look around.* Climate change has created conditions that make this not just good advice, but for firefighters, something that may well one day save their lives.

For people whose work turns around wildfire, the Dude Fire was a kind of starting gun in a race to better anticipate what climate change might do to the forests of the Southwest. What might happen, for example, to Arizona's critical summer monsoons—a lifeline for everything that lives here? The summer high-pressure cells that drift into southern Arizona are getting ever bigger as the climate warms, and that may be causing the monsoons to become more sporadic. Their failure to show in 2020 led to massive wildfires roaring across tens of thousands of acres south of the Mogollon Rim, where the land is dominated by several species of oak. Farther south still, in the Sonoran Desert, even that master of drought the saguaro cactus succumbed by the thousands, as some still are today.

~

If the Dude Fire was an early warning that wildfires were entering a new age of intensity and complexity, the supreme come-to-Jesus moment in this region was the 2002 Rodeo–Chediski Fire, which also burned through the ponderosa forest along the Mogollon Rim. It was made up of two human-caused fires that merged into a mind-blowing inferno; at 468,000 acres, it was, at the time, the biggest wildfire in Arizona history—more than four times the size of what today qualifies as a so-called megafire.

The Rodeo–Chediski Fire started running for the Mogollon Rim on the morning of June 19. The narrow canyons incised into the stony face of these cliffs tend to channel winds, and on that day they did exactly that. The canyons acted like giant bellows, pushing up flames until they finally broke over the rim in waves some six miles wide and more than 250 feet high. At one point, the fire ate up ten thousand acres of ponderosa in fifteen minutes.

The Rodeo–Chediski Fire called for a refiguring of the wildfire behavior playbook that firefighters had spent decades creating. This burn not only ended up moving faster than many thought was possible, but it also carried flames higher and hotter than most had ever seen. Even seasoned veterans of wildland firefighting were astonished. Managers did in fact overhaul their fire models; the new versions of those models suggested there would increasingly be times where rapid containment of a wildfire was impossible.

In the American West, for every degree of rise in temperature, the number of acres that burn will increase by about 400 percent. This means that if we manage to hit the target goal set in Paris during the 2015 climate talks—a goal our current government has abandoned— the area burned annually in the West will grow by 500 percent. That's about six million more acres every year, an area the size of Vermont. More homes, businesses, livestock, and pets will be lost. More people too.

∼

If any ecologist in the 1990s hoped to one day nurse the ponderosa forests back to something like they were before European contact, those hopes have long since vanished. Many researchers today don't even like using the word *restoration*, viewing it as unreasonably ambitious; what we can mostly do now, say an increasing number, is pick up the pieces. While we can make forests healthier by helping them get back to natural densities—thereby leaving each tree with a bigger share of water—as we grow warmer and drier, that share will increasingly be less than what's needed to stay alive. Trees will not only die of thirst but also, in their weakened state, become prey to insects and blight and root rot. Those that do survive and cast seeds will have another kind of problem, as conditions are often leaving seeds with little chance of sprouting. Finally, disappearing as well will be many of the creatures that made homes in the forest: nuthatches and owls and woodpeckers, tree frogs and leopard frogs, chipmunks and gopher snakes and tassel-eared squirrels.

Following the Rodeo–Chediski Fire, land agencies set a goal to "treat" three million acres of forest annually across the eleven western states, much of it ponderosa. This would involve a mix of mechanical thinning and prescribed burning, efforts meant to lower the density of the tree growth and to remove debris from the forest floor. There was every confidence such treatments would work; during the Rodeo–Chediski Fire, for example, stands of mixed-age ponderosa that had been thinned or intentionally burned survived mostly intact.

The tool of prescribed burning, while effective, is undoubtedly as much an art as a science—one best handled by highly experienced firefighters. This is especially true in an era of climate change, given how weather conditions are increasingly prone to shifting on a dime. The disastrous 2022 Calf Canyon/Hermits Peak Fire in New Mexico—the largest, most destructive wildfire in the state's history—was the result of two prescribed burns lit in April, fanned into infernos by sudden, unexpected winds. Understandably, the loss of hundreds of homes left a lot of people angered by the mere talk of prescribed burning. Following the Calf Canyon/Hermits Peak Fire, the Forest Service declared a three-month-long, nationwide moratorium. Again, though, the risk of this tool is dramatically lower when the job is in the hands of thoroughly seasoned firefighters.

Something that's proving harder to deal with is the hot-button issue of people breathing smoke from nearby burning of the forest. Naturally, the places that have the highest priority for prescribed burning are those where wildfire could threaten human life and property. Despite best efforts, though, that can sometimes mean days of bad air for people living nearby—not good news to anyone with respiratory issues. Then again, not burning the forest at all brings health risks of its own. Smoke from actual wildfires—big ones, like the massive Dixie Fire in California, which also burned through an unnaturally vulnerable forest—often spew bad air across the entire western United States, and beyond.

For all this, maybe the hardest thing to deal with is the sheer size of the task. To be effective, prescribed burning needs to be repeated every five to fifteen years, which translates into a staggering amount of money and labor. Across the eleven western states, we're currently treating about two of the three million acres called for in the wake of the Rodeo–Chediski Fire. That doesn't sound awful. At least it doesn't until you hold it up against the fact that a few years ago, federal fire managers took another look and decided to upgrade the goal from three million to eight million acres annually.

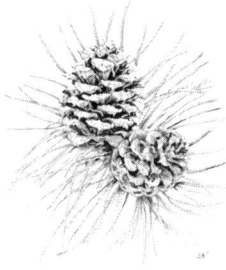

Chapter Nine

I stay all the next day and night on the Mogollon Rim, leaving early the following morning, heading north across the vast plateaus of ponderosa forest—some green, some standing black and gray in the grass of old fire scars. I need a break from chasing climate change. So I've decided to steer for a grove of ponderosa that's been special to me for forty years—a huddle of old friends. Friends that, shortly after first meeting them, went to work helping me mend something badly broken.

In the spring of 1981, my first wife and I were driving south from Idaho to west Texas, where I was to write a story about cactus thieves ripping off thousands of plants from Big Bend National Park. We'd just finished a month taking care of a ranch in Stanley, Idaho—shoveling snow at 30 below zero, feeding lodgepole pine to the woodstove in our cabin like it was the firebox of a locomotive. When we got out of the van for gas in Phoenix, it was 71 degrees. Flowers were everywhere. It was, to say the least, seductive. We both knew we'd probably never settle in a city for long, but damn, on that particular day this one looked good.

We made the move in early May, but barely a month in, things weren't going all that well. The heat of summer was on, with temperatures soaring to 115 degrees and more. Every Friday afternoon, we packed our van—the one with no air-conditioning—and headed north, climbing up and over the Mogollon Rim, there to replant our lives for the weekend in the cool, unruffled beauty of the ponderosa forest. At the risk of sounding dramatic, I can recall climbing out of the van and actually going weak in the knees—in part over the feel of that cool summer air, but also because of how it smelled—the soft vanilla scent of the trees mingled with the lemony pepper of needles lying on the forest floor.

On one of these escape weekends later in the summer, we happened across a place called the Mormon Lake Lodge. We pulled into the parking lot, shut off the engine, and then sat there for a minute looking at one another. We didn't have to say anything; we just knew the time had come. What did we have to lose by going in and asking?

We were the only customers, and we settled onto pinewood barstools and ordered a couple beers. The bartender was a big man, busy restocking bottles of Jim Beam and George Dickel after the Friday-night crowds. He was quiet but friendly, telling us about his crazy busy summer, what with lots of big crowds coming up to escape the heat in Phoenix. Finally, during a pause in the conversation, I blurted it out: "Would you know of anybody around here who might be looking for someone to take care of their place through the winter?"

He set down the two bottles of George Dickel he was about to shelve, wiped his massive hands on a bar rag, then turned to look us right in the eyes, head tilted, sizing us up. Apparently satisfied, he lifted a pen from a cup sitting on the bar and wrote down a name and a phone number. Joe Lockett, it said.

"He's got a line camp south of here. He might be looking for someone to keep an eye on it."

It was that easy. Like this big-handed, no-nonsense man had been waiting for us to walk into his bar.

"Who should we say told us to call?"
"Tell him Fat Jack."

⌒

Joe Lockett was a dyed-in-the-hide rancher. He stood five-nine, bow-legged and handsome with a black handlebar mustache, with ring marks from cans of chewing tobacco imprinted on the pockets of his jeans. He'd suggested we meet at Denny's in Flagstaff, where he spent the next two hours trying his best to hit the bottom of a bottomless coffee cup. We'd never seen anyone drink coffee like that. He told us that the next day he'd be riding fence line, and it was a lot of work, threading through the woods, dropping in and out of draws. Coffee kept him straight in the saddle.

"There's no electricity," he explained, referring to the place we'd be caretaking. "And once freeze-up comes in December, no running water. You'd have to use the spring tank, and that's about three hundred yards from the old cookhouse."

The cookhouse was the only building someone could live in, we learned, though not by much. We'd be sharing the space with mice and, in the attic, a packrat or two. The nearest neighbor—not to mention the nearest plowed road—was seven miles away; we'd have to have a snow machine to get back and forth. In exchange for watching the place we could live rent-free. He'd provide us with firewood, and that would be our only source of heat.

It sounded perfect.

We lived in that little cookhouse for seven months, fall to spring. Breakfast was oatmeal and powdered milk, and dinners cornbread and chili and pots of spaghetti heated on the sheet-iron woodstove. Left-overs never went bad, in part because the whole place was a refrigerator. When we went outdoors—and we went out many hours every day, either walking or skiing—it was straight into the surrounding ponderosa forest, which also here and there cradled small patches of quaking aspen, their trunks the color of last week's snow.

Jane worked on a series of kids' books she was writing. I sat at a plank table with an old manual typewriter, putting together ideas for articles to send to magazine editors. Once a week we loaded up those pages, tossed our laundry into a wooden sled I'd cobbled together from stray boards around the cookhouse, and headed off by snow machine through those seven miles of frozen forest to the van; then, shovels in hand, we'd dig out of the snow.

In Flagstaff we'd found an elderly lady willing to rent us a room with an electric outlet for forty dollars a month, where I retyped every query letter and article on a spiffy blue Royal electric typewriter my parents had given me the previous Christmas. At the end of the day we'd drop the queries—and later, actual manuscripts—off at the post office and go to do the laundry and drink beer at the Flagstaff Suds and Duds. And if we were feeling flush, maybe grab a burrito at La Fonda. By the time we drove back to the parking lot and fired up the snow-mobile it was long past dark. Sometimes the moon was out, bright enough to make the snow shimmer, strewing it with long tosses of blue diamonds.

~

There was a catch, though, in this otherwise sublime adventure. A big, heart-busting sadness. On October 8, as we loaded the van in Phoenix for the long drive to Lockett Ranch, my mother called, her voice so tight I could barely tell it was her. My dad, a sheet-metal worker, had been working on the roof of a nine-story courthouse in Plymouth, Indiana. Temperatures the night before had dipped below freezing, and when workers showed up the following morning it was decided someone from the crew should test the roof for frost, make sure it was safe to work on. My dad volunteered. Despite it being a law that the construction company have safety lines on the roof, none were ever installed.

Now my mother was telling me he'd fallen ninety feet and was broken to pieces. We caught the first flight back, rushed to the hospital where he lay with his forty broken bones, heavily sedated, a raft of machines beeping and pumping. At one point he came slightly alert and thought he was late for work; struggling to get out of bed and get dressed, he had to be held back. Three days later, a blood clot from somewhere in his bleeding body cut loose and stopped his heart.

For months after that, back at the Lockett Ranch, when I wanted to talk with him, or more often just cry over him being gone, I went out into the ponderosa forest. It was there I thanked him for his patience—my god, he was a patient man. I told him how I should've believed him at eleven, when he stood in the doorway of my bedroom and told me that if there was ever something I wanted to change about myself, it would happen slowly, one small step at a time. Out in that snowy forest, I found myself wishing I'd found the courage to tell him about the things that made me afraid. Maybe even ask him if he had a list like that too.

A mile or so from the ranch was a beautiful mature ponderosa tree, and unlike most in the forest, it had kept a few of its lower branches. This made it easy to climb. About six weeks after my dad's death, just a few days before Thanksgiving, I put on my snowshoes and set out through a foot of new snow, heading for that tree. I spent some time tamping down the snow to better support my weight, then kicked off the snowshoes and hauled myself up onto one of the lower limbs. From there I hoisted myself up to the next level, and to the one above that, finally stopping about twenty feet off the ground. I don't know how long I stayed, but it was quite a while—sitting shoulder to trunk, blinking into the blue sky. There was no wind, and the forest wore the kind of quiet that only comes in the depth of winter. One minute I was smiling, imagining my dad down at the base of the tree, squinting up at me, warning me away from stepping on dead branches. The next minute I had tears running down my face. Tears for my wilted, broken mother. For my brother, too, who on first setting eyes on our dad in the hospital, wrecked and helpless, morphed before my eyes into a scared little boy again. And tears for me, too, thinking how to him I must've looked the same.

I went to that tree several more times that winter, always calling it by my father's name, Bob. I came to know a pair of squirrels there who couldn't stop chasing one another—sometimes for fun, other times as part of some squabble. Sometimes the brown creepers or the chickadees or even the sapsuckers hung out with me, quietly picking at the bark while I watched. At the base of the tree were tracks in the snow from white-tailed deer and red foxes—and once, the footsteps of a mountain lion. I came to know the sound of the breeze in the tree's needles, how it changed some depending on the direction of the wind.

Somewhere in that winter I'd read about the East Indian scientist Jagadish Chandra Bose, a Cambridge-educated biologist in the late 1800s, member of the Royal Society, who believed plants could feel pain, even partake in emotions like affection. "Everything in man," he said, "has been foreshadowed by the plant." It was thanks to Bose that on especially sad days in the pine forest, I got to wondering if Bob the ponderosa might well know that there was a human sitting in her branches, one badly in need of a good tree.

When we die, no matter if our bodies are buried or scattered as ashes, sooner or later we go back to soil. And from there, many pieces of what was once us will be gathered by trees. I think of my father like that. How maybe some measure of Bob molecules was taken up by the line of big, chesty oaks that frame his grave. A thumbnail story of his life might go something like this: A shy and open-hearted small-town kid from Indiana grows up to be a blue-collar working man. Then a husband, and soon afterward, a father. A good father. He builds a cottage in the woods with his own hands. He listens to my mother playing the piano and singing *Smoke Gets In Your Eyes*, and when she finishes, he tells her how nice it was. And then one day, long after his death, he finds his way into a tree. He becomes heartwood.

$\sim$

I park for the night a few hundred yards from Bob the ponderosa. A pair of red-breasted nuthatches are flitting about—feisty, tightly packed little birds climbing up and down the tree trunks, nabbing insects, tossing out their tin-horn calls like they're in charge of, well, everything. A monogamous couple, they've been working together all summer—beginning with building a fine nest, lining the entrance with a sticky smear of ponderosa sap, possibly to deter would-be egg robbers. They've chased away everything from trespassing woodpeckers many times their size, to warblers and wrens. They hatched their young, fed them, and watched them fledge. And before long—far sooner than most other birds—they'll head south, dropping off the Mogollon Rim, arriving by late August onto winter range.

For all their bird smarts, the nuthatches here have been dwindling over the past hundred years—in part due to logging and grazing, but more because of fire suppression and its legacy of smaller, more crowded trees. It's birds like these, who either feed in the canopy or make their livings cruising up and down the trunks, that are faring worst in the ailing ponderosa forests: mountain chickadees and white-breasted nuthatches, pygmy nuthatches, brown creepers, bluebirds and warblers, swallows and song sparrows, nighthawks and swifts.

As the stars begin to wink on, I head off for a walk. A flammulated owl—barely bigger than a robin, and so very secretive—is tossing out calls that sound soft and flat, like she's using her library voice. As I sleep tonight she'll be working, plucking beetles off the ponderosa, grabbing the occasional moth. Back in the Mogollon Mountains of New Mexico, there was that great horned owl. And now this flammulated. When I first started living in the West in my early twenties, I met an old hunting guide who said that seeing owls meant it was time to be alert. That your next best move would come by focusing.

Tonight I dream of my older brother and me as kids in Indiana, in winter out on our sleds. He's warning me to be careful on the way down, be alert so I don't end up running into a tree, which in real life, I'm embarrassed to say, was something that actually happened, and more than once. In the dream I start down the hill on my sled, watching everything carefully, ready to roll off if any trees loom—excited but scared. I can't really say if I made it down okay. Partway down that dream hill, I'm startled awake by a mule deer outside the back window of the van, letting out a big snort—the sound deer make when telling other deer to pay attention.

A clear morning in the ponderosa forest is the epitome of my happy place. The slant of early sun feels reassuring—almost spiritual, though not in a hallelujah kind of way. At its best, more the kind of place where I'd hope to end up after my last breath. A beautiful forest like this, washed in this very light, maybe with some wise being, there to help me look back at my life—the hopes I had, the choices I made. Being given in that moment, in those woods, some chance for reconciliation.

Right now, though, peering deep into these the woods, I can almost imagine the whole grove blurring into one of those ancient forests of metamorphosis, where in the dim light one thing can suddenly turn into another. Where something is always moving in the shadows, carrying a presence but impossible to fix in place. Soon it feels like the whole woods is wrapped in it: every needle and root and reach of cambium, going on about the business of life, tweaking and adjusting minute by minute, trying always to thrive in a world where nothing ever stays the same.

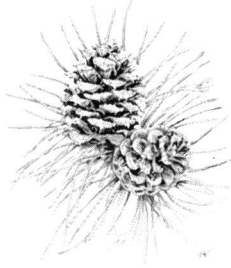

Chapter Ten

M uch as was true in New Mexico, here in Arizona some of
the ponderosa forests that have burned or been killed by
beetles likely won't be coming back. In one large survey
of the Arizona ponderosa groves that have burned since 1995, fully a
quarter are no longer growing any trees at all, covered instead with
grasses and fire-tolerant shrubs. Some of the trees lost stood just to the
north of the Mogollon Rim. Others are near Kendrick Peak, northwest of Flagstaff, or just beyond the feet of the San Francisco Peaks.
The losses are bigger in the so-called Arizona Strip—that swath of
land between the north rim of the Grand Canyon and the Utah state
line. Were you able to look down on things from up high, you'd likely
conclude that the world's biggest ponderosa forest is starting to fray.

My sense of how hard it will be to mount a comeback for these trees
is only heightened by how hot and dry the days are now. I've driven
barely thirty miles from the Bob tree, intending to spend a few days
exploring the forests closer to Flagstaff; it's just now noon, and already
it's 95 degrees. I can practically smell the fire danger, a thin gruel of
terpenes simmering in the bone-dry air—almost souring them, like
white wine going to vinegar. On some trees, needles are already going

brown. It's only mid-July, and yet sitting outdoors in the afternoons, the forest floor is so tinder-dry I can hear the footfalls of mice. As for that famous Southwestern light, it grows weaker with every passing day, shrouded with smoke from distant fires.

Like other trees, ponderosa release small amounts of carbon dioxide, mostly at night, through the small pores in their needles. When those pores are open, precious moisture sucked up from the roots into the branches and needles evaporates into the air. If those roots run low on water, and if the rest of the tree keeps sucking like a kid with a milkshake straw, the channels that carry the water may collapse from the pressure. It's known as cavitation, and when too much of it happens, the tree dies. These feel like cavitation days.

South of here, both Phoenix and Tucson are about to break their longest streak of consecutive days with temperatures above 100 degrees; the old record of seventy-six days in Phoenix will this year swell to a mindboggling 113. Show Low, Prescott, Winslow—all are setting records. And a few days from now, on July 22, this beautiful planet will experience its hottest day in history.

~

With the heat and the thirst crackling on, I find myself thinking more and more about how we ended up in such dire straits. Yes, much of the fault for these fading forests can be traced to our having insisted on suppressing wildfires. And it most certainly has to do with the blithe and boozy way we've clung fast to the burning of fossil fuels, and also the way we keep cutting down the world's forests to make room for grazing cattle.

But those are end results. What is it that drove us to choose to play so fast and loose in the first place? Why do we have so much trouble aligning ourselves with the deepest resiliency wisdom of this planet, offered up by every forest, which is that individual lives are made powerful through community? Why are we still reluctant to face our predicament—not just so we can solve it, but first to see if we can together, as poet Stanley Kunitz put it, reconcile our hearts to this feast of losses?

Many of the key decisions that caused this pot to boil can be traced back a good four hundred years—to that brilliant and bedeviling era known as the Enlightenment. The era celebrated for having given birth to modern science. And in the process fueling a fiercely rational, mechanistic style of inquiry into the workings of the physical universe; one so successful it would bust out of the confines of physics and chemistry, where it was incredibly useful, to blow across every landscape of our lives. This effort to apply the tools of early science to life wholesale would be our undoing, channeling the way we think, limiting the breadth of how we see ourselves in the world.

Of course, no system of thought shows up out of nowhere, and the precepts of early modern science are no exception. Everyone from the Greeks to the Chinese, the Sumerians to the ancient Egyptians, as well as countless Indigenous peoples, had long before been doing what the Enlightenment took to a new level: building a sophisticated knowledge of how things happen in the physical world, and then putting that understanding to work to either alter the present or anticipate the future. Yet the Enlightenment—and for these purposes I'm broadening the era some, placing it from the 1630s to the 1780s—used its powers of inquiry to forge a thoroughly human-centered knowledge. One eager to dissect the universe in a controlled, mathematically sophisticated, and—by its own admission—often ruthless manner.

The boys of the Enlightenment Club (and it was most assuredly a club created by and for white men) believed one could uncover the whole of reality by breaking things down and then rationally studying the parts. Learning what made the universe tick, said everyone from René Descartes to Isaac Newton to Francis Bacon, was little different than discerning how a clock worked: to figure it out, you just needed to take it apart. And when you did, you'd eventually find the part of the machine that drives it. If we stayed committed to detached, objective investigation, if we stayed the course in disassembling the machine, eventually we'd be able to control, well, almost everything.

Lest you doubt the lingering power of such an idea, consider that just twenty years ago some very smart people were touting the promise of genetic engineering for giving us the ability to move genes from one place to another, mixing and matching, replacing faulty ones much as you might change out a broken gear in a car transmission. While it's true we've learned some very useful things about gene splicing and editing, the thought that a single gene is solely responsible for a specific behavioral trait, or even a particular disease, ignored the essential role of the environment. How a gene expresses itself—be it in a tree or a honeybee or a bacterial cell—is determined by countless interactions with the wider world. The leading edge of life is relationship.

Early modern scientists favored a clever kind of inquiry known as the scientific method. For something to be declared true it had to be reliably observable—either directly, or from conclusions made through repeatable experiments. The scientific method involved isolating an object or a process, making an educated guess about a certain quality it might have, then designing an experiment to see if you were right. Whatever you decided to study, though, it would need as much as possible to be isolated. Investigation of the world became a matter of subject-object thinking: The scientist was the subject—the rational observer—and the thing being studied (or, later, even the person being studied) was the object. Descartes called this the "spectator view," and it worked incredibly well for solving lots of puzzles, especially those in physics and chemistry.

Make no mistake, these rules on inquiry were often practiced in an aggressive, take-no-prisoners kind of way. A common opinion of the early Enlightenment was that all of nature should be forced to yield to our inquiry, and then to our control. As Francis Bacon put it, "We must use the rack, and force nature, as we force a witness, to give her evidence." And again, it wasn't long before every discipline in the Western world, many of them way outside the physical sciences, sought legitimacy by pledging themselves to this same aggressive quest for proof-based reality.

These were brilliant men, right about things often enough that it was easy to imagine them as being right about all of it. Even their truly bad ideas had remarkable staying power. Many, for example, thought animals not only lacked emotion but were incapable of feeling pain. Descartes described them as automatons, as limited and predetermined as any machine—a belief that allowed him and others to do things like cut open live animals with no concern. Incredibly, it would be 250 years before most people accepted the idea that animals were sentient enough to feel all kinds of things, from pain and suffering to excitement and joy. Even then, some resisted such notions—in large part because emotions and feelings couldn't be decisively proven with the scientific method; any speculation about things unprovable was kryptonite, dangerously likely to pollute our understanding of reality. In the 1960s, when Jane Goodall first spoke to her colleagues at Cambridge University about the chimps of Gombe Stream in Tanzania, noting the rich range of personality and emotion they displayed, many of her colleagues ridiculed her outright, insisting that chimpanzees weren't capable of either one.

This rigid objectification of nature supercharged a system started long before the Enlightenment, where each life form was compared to other members of its kind or to other species, and then ranked. People were the most advanced beings in all creation; yet even within that group, there was a best and worst. It's telling that across the past four hundred years, those humans considered "less than," those that would end up most ruthlessly dominated and abused—girls and women, people of color, Indigenous cultures—were without exception portrayed as being "closer to nature." It wasn't a compliment. Like nature itself, they were said to be inferior, lacking in both aptitude and reason. This is how we end up with nonsense like that expressed in the late nineteenth century by noted anthropologist Herbert Spencer, who stated matter-of-factly that women's brains simply weren't as evolved as men's. Or in 1900, when Paul Julius Möbius, whom Sigmund Freud called a father of psychotherapy, published "On the Physiological Feeble-Mindedness of Women." Among those in power, such declarations—and there were lots of them—barely raised an eyebrow.

~

Dissecting life to figure out the machine, to simplify it, became a big thing. As modern philosopher David Graeber describes the social sciences even today, we engage "in a game of make-believe in which we pretend, just for the sake of argument, that there's just one thing going on: we reduce everything to a cartoon so as to be able to detect patterns that would be otherwise invisible." Graeber acknowledged that such simplification can be a critical first step in discovering something new. "The problem comes when, long after the discovery has been made, people continue to simplify."

Probably no one summed up this mistake better than the brilliant British philosopher Mary Midgley. Ways of thinking that were enormously worthwhile in the early physical sciences, she said, "have been idealized, stereotyped and treated as the only possible forms for rational thought across the whole range of our knowledge." The trouble lies not in the methods, she added, "which are excellent in their own sphere." Rather it's "in the sweepingness, the dramatic zing, the naïve academic imperialism that insists on exporting them to all sorts of other topics."

Topics like economics. Again, given how complicated social systems are, economists have understandably had to play the game David Graeber talked about, pretending that only a few key things are going on. Their results are often as elegant as a good math equation. And at the same time, they can be disastrously misleading or incomplete. Especially when the sole goal is profit.

It was long thought to be cheaper, and therefore more efficient, for coal-fired factories not to have scrubbers on their smokestacks—which is why, in the US, none were used until some forty years after their invention. Accounting for secondary "costs," like the damage to people's health, or related losses in worker productivity, was too messy. Likewise, while the nineteenth-century textile mills in the Northeast were never easy places to work, as they became more driven by the economics of efficiency, they became a living hell for workers. By the turn of the twentieth century, it was considered more efficient to use up teenage girls in the shirtwaist factories, and replace them with another crop of young immigrants when they sickened or died of tuberculosis, than it was to spend money creating safer workplaces with better air. Similarly, for decades the most efficient way to get rid of toxic waste was to dump it into waterways adjacent to the factories—which again offered all the farsightedness of a mole, ignoring the cost of lost fisheries, increased rates of cancer, and widespread brain and other neural injury to children. It's telling that health and labor advocates only became truly effective when they learned to speak the language of money—when they were able to work out the costs in dollars and cents from disease and degraded health, missed work, and premature death.

Sometimes arguing the case for the Earth in economic efficiency terms—and I, too, have done it countless times—leads to plain weirdness. Like the time a team of nature-friendly economists in the 1970s set out to determine the dollar value of a day spent in nature. Being good social scientists, they churned out enough graphs and tables to satisfy the most demanding mavens of precision measurement; in the end they pronounced such days to hold the equivalent value of a movie ticket, which back then was about five dollars. By multiplying that five-dollar figure by the number of people recreating in a particular wild place, conservationists were sometimes able to show that the land in question held more economic value from people enjoying it than it did from being given over to wholesale logging or mining.

Likewise, when it came to large-scale mining projects in the West, especially gold mining, we conservationists learned to tout the very real cost to taxpayers of paying for cleanup when companies abandoned projects, including the nightmare damages that came from leaking cyanide leaching ponds. Elsewhere, we helped preserve places by unveiling research showing that undeveloped public lands increased the worth of real estate in nearby towns and cities—which they really did. This argument worked because it held the proper homage to profit, giving politicians the cover they needed to cast their vote for land protection.

With economics having the final say as to whether something should or shouldn't happen, we avoid any need to have those squishy, imprecise conversations about sense of place, about whether nature should have legal or moral standing, about what ethically we might owe future life on Earth—which are exactly the weed patches Descartes dismissed as "the enemies of rational thought." Without question, lots of wild places have been saved by economic arguments. Hopefully more will be in the years to come. But there's something sad, bordering on tragic, about the fact that, as Robert Pogue Harrison described it decades ago, when it comes to nature "the language of utility is the only language that has a right to speak."

Science, claimed renowned psychologist B. F. Skinner, is obligated "to restrict itself to what we can see and what can be manipulated and measured in the laboratory." Affinity and empathy and compassion for life on Earth might exist, but since it can't be objectively measured, it's irrelevant. Instead of making room for those essential values, seeing them through an expanded sensibility, we've reduced them to occasional niceties, topics worth an op-ed or two following a big oil spill, or maybe as something to talk about on Earth Day. But here's the thing: For all the technological brilliance we bring to solving a problem like climate change, in the end the work depends every bit as much on those squishy, imprecise matters of the heart.

～

Given how the fundamentals of early science spread to almost every human endeavor, it would have been strange if forestry hadn't followed suit. In fact, forestry was arguably the first nature-focused discipline to arise in the wake of Enlightenment science. In the early 1700s, especially in Germany and France, a group of properly rational men turned to the tenets of science to help their countries recover from what had been long-standing abuses in the forests. Being good Cartesians, they had no intention of returning the forest to its natural state. Far from it; the idea was to make them better. Many explained this was exactly what God intended us to do. And, again, a lot of the work to make the forest better was about making it more efficient—which meant more profitable. Starting back in the 1700s and going strong in a great many places to this day was the immediate and long-term goal of increasing the capacity of the land to crank out commercially valuable trees in the shortest amount of time possible.

Notably, this early no-nonsense approach to the forests, the determination to make them into crops, dovetailed well with thoughts about the woods being laid down by the church. Pre-Christian notions of wild forests rested in seeing them as places of fuzzy edges, where one thing tends to blur into another—a land of flux—qualities reflected in no end of early story and myth. As Robert Pogue Harrison described the wild forest, "it appears to us to have no irreducible distinctions—no noise that does not sound like a response to some other noise, no tree that does not fuse into the arboreal confusion." But for a church no less fond of order than science was, any world this hard to nail down was a hotbed for evil. Indeed, one of the creatures living in those shadows, the wolf, was soon declared to quite literally be "the devil's dog."

The wild woods, as some Christian leaders put it, mocked the orderliness intended by God. While the church continued to have problems with science when it contradicted scripture, they were more than happy to have scientists deconstruct the wild woods, stripping the forest down to something knowable and controllable; clean up the mystery and make something useful. Descartes, himself a committed Christian, in one moment both posed and then answered a philosophical question of the day in a way that satisfied nearly everyone: "How does one walk in a straight line through the forest? By methodical deforestation."

Not surprisingly, it was soon determined that the creatures of the forest would also need management. Which mostly meant protecting the ones we liked and slaughtering the ones we didn't. Deer were good. Foxes and jackals and most other predators were bad, in part because they ate the good wild animals we wanted for ourselves. Forest improvement would mean keeping the good and wiping out the bad. For some early foresters who happened to be Christians, I imagine them thinking this work was a sweet spot: a job offering the chance to improve nature, protect the good animals of the Earth while getting rid of the bad ones, and, as icing on the cake, derail the devil.

Many European fears about the unkempt forest would cross the Atlantic to land in the New World. But in addition to wolves and foxes and owls and snakes and other loathsome creatures, the Puritans also encountered Indigenous cultures, which again they very much thought of—in the worst way possible—as being "closer to nature." To many European settlers, the religious practices of Native Americans were flat-out witchcraft. Anglican Bishop John Jewell wildly claimed that Native people drank blood and regularly sacrificed boys and girls "to certain familiar devils." The future Archbishop of Canterbury described the Indigenous people of northeastern America as regularly engaging in incest, sodomy, witchcraft, and cannibalism. And heroic Captain John Smith, whose hide was saved by Pocahontas from certain execution at the hands of the Powhatan, referred to the forest tribes as a cult of devil worshippers.

~

By the time forestry took hold in America at the very end of the 1800s, the devil had mostly yielded the stage to science. (That said, the American West today still has over seven hundred place names referring to the devil or demons.) But while science may have ostensibly pushed on without religion, it stayed firmly wedded to the idea that the woods should be improved by maximizing efficiency, which was understood as commercial productivity. We were a growing nation, argued the American foresters of the late nineteenth century, in need of wood. It was past time to get busy.

The future head of the Forest Service, Gifford Pinchot, claimed the American foresters who had come before him were too careful, that "they saw too many lions in the path." He complained about these men thinking it was necessary to know the biology and life history of all that was growing in a forest, and only then, armed with that considerable understanding, hatch a management plan. To Pinchot, that was overcautious. A waste of valuable time. As he often liked to say, we were at best twenty years out from a timber famine. And when Pinchot said things like that, people listened. As Stanley Horn recalled in a 1940 issue of *The Atlantic*, at the turn of the century "there was hardly a newspaper reader who did not believe that by 1927, or 1932 at the latest, a tree would be just about as rare in the United States as in Iceland or China."

You might think such harvest-driven zealotry would be enormously reassuring to the timber industry. But timber companies were skeptical, afraid of being forced by the government not only to cut more selectively but also to replant the trees. And replanting hardly fit with a saying common at the time among timbermen: "Cut out and get out." The forest was seen as the equivalent of a mine, observed Stanley Horn, "to be worked out and abandoned when its merchantable material was exhausted." This fresh crop of science-educated foresters, on the other hand, were fine with working the forest hard—just not to the point of totally exhausting it. Walking that fine line would take scientific management. So Pinchot and some of his fellow foresters set sail for Germany, there to learn at the feet of the Cartesian virtuosos.

The lessons that excited Pinchot and company while they were in Europe turned on how the Germans—and, to no small degree, the French—had managed to turn wild forests into sustainable crops. The European foresters taught them about harvesting and replanting

schedules, about the best ways for thinning unwanted trees, for dealing with tree-killing insects, for fine-tuning growth rates. All this they
carried back across the Atlantic and out to the American West, where
it would guide management of the nation's forests for the next eighty
years. Leafing through historical records of national forest operations
in the first half of the twentieth century is like drinking beer with Pinchot in a pub owned by Descartes.

Given how often the ponderosa forest was front and center in these
harvest plans, I don't have to look far to find evidence of it around
Flagstaff. One place it shows up is in the notes of a truly accomplished
forester in the first half of the twentieth century named Gus Pearson—
a man chosen by Pinchot himself to establish a forestry research station here, which came to be known as the Fort Valley Experimental
Forest. The Fort Valley Forest still exists today, an important setting
for researchers to better understand how forests are adapting to climate
change.

As for Gus Pearson, he spent a lot of time living alone in a tiny
uninsulated cabin in the woods, apparently meeting most of his social
needs through conversations with his mules, Pat and Mike. By the time
he retired in 1944, it's safe to say he knew more about the workings of
ponderosa pine forests than anyone on the planet. A 114-page mono-
graph he authored in 1950 outlines in meticulous detail every imaginable aspect of the tree. Yet like all his writings, this one was done with a
steady eye to rendering the forest as commercially valuable as possible.
As Pearson himself described it, everything he did assumed "a policy
which looks toward growing maximum timber crops."

Pearson was neither reckless nor cavalier. He acknowledged the rec-
reational and aesthetic values of the woods, writing—seemingly with
some relief—that "foresters no longer believe that every acre of land
that can be made to grow timber must be used for that purpose." Still,
he was hired to treat the forest like an always-solvent, sustainable bank
that doled out currency in board feet of lumber. Little wonder that he
and his team were ecstatic when they figured out how to shorten the

cutting cycle from about sixty years, which was common in the days of railroad logging, to just ten or twenty. "The old rule that a minimum cut of about 5,000 to 6,000 board feet per acre is necessary for economic justification of logging operations," he wrote, "no longer blocks effective sustained yield."

In this world, virgin stands of ponderosa were "over-mature," a label having to do with the fact that after a certain age the quality of the wood in a ponderosa tree begins to decline. The ideal forest was the one with few or no trees over two hundred years old. Likewise, a major objective of silviculture in these "near-desert forests," as Pearson called them, "must be to assist nature in making the necessary adjustments to moisture conditions, thereby forestalling stagnation of growth and the appalling waste which is inevitable under the law of 'survival of the fittest'—a phrase which from the standpoint of lumber production means survival of the unfittest."

Instead of fifty board feet per acre, Gus concluded, it was possible to grow 150 to 200 board feet per acre. Instead of a log crop in which 90 percent of the lumber fell in the lower quality grades, and only 10 percent higher quality, good silviculture could reverse that. "The major problem," he said, "is no longer how to harvest old timber, but how to grow new crops which should become available 50, 100, or 200 years hence."

Just as had been the practice in early European forestry, any creature that threatened harvest targets was something to be "carefully regulated to prevent serious damage to the major resource." Tassel-eared squirrels were bad because they cut pinecones and ate the seeds that could be growing the next crop of trees—which is why from 1940 to 1944, in this one experimental forest alone, just under seven hundred were shot out of the trees.

And then there were porcupines, which called for poison: one part strychnine to sixteen parts salt, mixed with melted lard or bacon grease to form a paste, placed in a bored-out wooden block on a horizontal branch above the reach of cattle. In the Fort Valley Experimental

Forest "as many as three dead porcupines have been found under a single bait tree." Then again, if you wanted to be doubly sure of solving the problem, long-term control "[could] be obtained by poisoning and shooting as complimentary measures."

The Forest Service was not then, and is not today, a cadre of heartless villains. Far from it. The legacy of Cartesian science—that pledge to make nature more productive—is a very big, very deep river that still flows through all of us. When Pearson wrote his tome on ponderosa management, the United States was flexing its muscles against unmanaged nature on every front: more corn per acre, fatter cattle and pigs, chickens with bigger breasts and able to lay more eggs. The US Bureau of Reclamation, steeped in engineering science, now and then championed the idea of big dams on the argument that the western rivers would otherwise "go to waste." In one year alone, the agency was given a green light for the Navajo, Glen Canyon, Crystal, Flaming Gorge, Blue Mesa, and Morrow Point dams, entrenching a kind of can-do conservation that favors storing things up over using less. At the same time, state and federal wildlife agencies sanctioned the slaughter of a whole host of birds and mammals deemed inconvenient, from foxes to coyotes to mountain lions, hawks to owls. In Pennsylvania alone, during roughly fifteen years from 1948 to 1962 some two thousand great horned owls were killed. And the people shooting them got five dollars for every bird.

Sometimes humans stepping in to make nature more productive is useful, reasonably benign. This includes our use of fire over thousands of years to promote healthy grasslands and forests, or things like adding cattle dung to farming plots to increase nutrients in the soil. But when we bring to the table an unwarranted confidence in our assumptions, we invite a whole host of unintended consequences.

If we're ever to heal life on Earth, Carl Jung said, we first need to get everything back into connection with everything else. He went on to say this meant resisting the trap of our intellectual assumptions. "We need to understand that we cannot *only* understand."

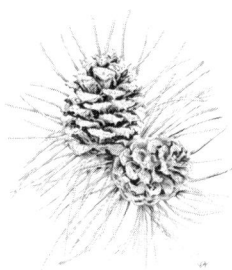

Chapter Eleven

If there's any place that can scour away the funk of patriarchal science, of having too often insisted that nature bend to our will, it's the one region of the West where such ideas seem most preposterous: the chaotic, heart-searing canyonscapes of southern Utah. I arrive here three days after leaving the sprawling ponderosa forests of northern Arizona, traveling the entire way under cloudless skies and searing heat. Southern Utah is ponderosa country too. But here the tree is more slotted into specific life zones, assembling into full-blown forests only on the high plateaus and on the upper flanks of mountains. Down low in the drylands it still persists, but often as lone trees rising from the bottom of random slot canyons.

Nowhere does nature seem more disinclined to fit the boxes we've built for it than here in canyon country. On high perches from the Colorado border nearly to Nevada, the landscape often appears as a fantastical maze, a giant version of the Labyrinth of Crete. On a clear morning the horizons can seem impossibly far away, as if their edges might be tilting into some other day. All in all it's a precious invitation to let your imagination break free and roam. Strange feelings can come on in places like this, sometimes at once delightful and disturbing. Like the one described by UCLA psychiatrist Mark Goulston, when he first stood on the edge of the Grand Canyon. "It took a fair amount of restraint to prevent me from jumping into it," he said, "because I was certain I could fly."

Here even time itself lies out in ways that break the brain. The language turns on what geologists call deep time. Or as I sometimes call it, blow-your-freakin'-mind time: how long it takes for an ancient ocean to become a desert. How long for water to carve a flat rocky plain into a mile-deep canyon. Or for a whole mountain range to push up into the sky and then crumble away again. Things measured in tens and hundreds of millions, and then billions, of years.

I grab a bite to eat in Torrey, then head off to put in a few miles on the trail in Capitol Reef National Park. For all the stupefying grandeur available from the high perches of canyon country, the path I end up on, winding through deep ravines, reminds me that down low the awe comes too from intimacy: a tiny ribbon of water running down the face of a shady rock wall, giving rise to a hanging garden, and at the base, pocket-sized patches of lilies and ferns. Twists of primrose unfurling in the bottoms of the washes across the bare sand, joined here and there by a smattering of yarrow and sego lily.

The heat's rising fast, though, and by the time I turn around to head back to the trailhead I can feel the squeeze. The return walk consists of me hopping from shade to shade like a lizard. Given the forecast for the rest of the afternoon, once back at the van I make the decision to go up in elevation. And just like that, forty minutes later I'm parking at nine thousand feet, high above Capitol Reef on Thousand Lake Mountain. When I check my weather app I see that the town of Torrey, sitting at the base of the mountain, has hit 108 degrees.

For someone out on the trail in such searing heat, especially down low, the proper course of action is no action at all. Simply to stop. To be still as long as it takes, napping in the shade or writing or singing or staring into space. I think this forced, nonnegotiable pause is actually a part of how true canyon explorers come to bond with this place. When you're on foot in the wilds and things get blazingly hot, it can become an invitation to pay close attention to what the day is asking of you. A time to go down with the birds as they get quiet, to pull yourself in like the primrose drawing in its flowers against the heat; to follow the mule deer and the rabbits as they retreat into the shadows. The most experienced canyon-country travelers meet the heat of the day not with a grit-your-teeth endurance but with a willing surrender, an acknowledgment that there's simply no other way to survive as a child of the high desert sun. Ground temperatures under a large ponderosa, or likewise in the shadow of a cliff face, are often more than twenty degrees cooler than the surrounding air. And so there you stay for a while, reading or napping, making memories out of doing absolutely nothing. Shade, as poet MaKshya Tolbert describes it, becomes a place.

When I read about long-distance summer travelers in the South-west, the first thing I think about is their dance with heat. Travelers like Francisco Garcés in the mid-1700s, happily wandering well south of here in even hotter conditions, from the Sonoran Desert to the Mojave—a man who never spoke of conquering the landscape but rather finding peace in agreeing to whatever it asked of him.

Or George C. Fraser, who in 1914 took leave from his successful Manhattan law firm, climbed aboard a train with his seventeen-year-old son, and set off to explore the outback of canyon country. In the years that followed, father and son made seven more extended treks, some of them in big heat, traveling by horseback along little-used trails, up into the ponderosa forest and then down into the bellies of the big canyons.

And, of course, Edward Abbey. That big, brilliantly irascible no-mad who didn't just wander this wild country but fused with it: with the ponderosa trees and the canyon walls stained with desert varnish; with the small depressions in the slickrock—*tinajas*—holding precious water long after the rains had come and gone; with the thorns and lizards and rattlesnakes and ravens. A man who came away wishing for his readers that their trails would always be crooked, lonesome, and dangerous.

But to me, and in fact for so many who love canyon country, the best poster child for wandering this world of stone and sky is Everett Ruess, a restless sixteen-year-old from Oakland, California, who first arrived in 1931. Everett was born hardwired for the vagabond life. To keep him from ambling off as a toddler, his mother tied him to a tree or, if he was inside, to a chair. At three years old, he managed to give his parents the slip and walk more than a mile from his home before finally being picked up by the police. Nor was he reluctant to head off at night—a habit that eventually led his puzzled parents to start putting him to bed with his feet tied together. (These being the same feet he later gave names to, calling one Jupiter and the other Jerry.)

When he was eight, Everett's family lived for a time in Valparaiso, Indiana, where he got his first chance to wander alone in a small toss of unkempt forest. He couldn't get enough of it: the bugs, the birds, the foxes, the deer, the racoons—the arrowheads he found lying about in the woods. All of it sparked a curiosity that would burn hot and bright for the rest of his short life. At sixteen, he began hitchhiking up and down the California coast, as well as making the occasional meander through the backcountry of Yosemite. And soon afterward, he thumbed his way to Arizona and New Mexico, then to canyon country—finding on his arrival one of the biggest sprawls of wild country left in the continental United States. If you were a teen nature geek like Everett—someone able to pay attention and learn from your surroundings, blessed with a reasonable measure of humility—you could be enormously happy for years rambling about in the American Southwest.

Some of Everett's precious obsession is captured in the journals he kept, as well as in letters home to his family. Starting in his early teens, he'd struggled with serious mood swings, and some of that darkness shows up now and then in his writing. Yet most of the pages hold an enviable, boundless eagerness—the joy of a person who knows with every fiber of his being that he's exactly where he needs to be. The canyon country became his North Star, and he was more than happy to pay homage to it on foot, mile after mile, year after year.

"I'll never stop wandering," he said in his journal. "And when the time comes to die, I shall go on some last wilderness trip, to a place I have known and loved. I shall not return."

Everett also channeled his fondness for these landscapes into his art, mostly through paintings, drawings, and woodblock prints. His mother, herself a talented painter, had a ferocious belief in the value of exposing Everett and his brother to art at an early age, teaching them watercolor techniques when they were still little children. In early elementary school, with the family living temporarily in New Jersey, his parents enrolled him in a prominent art school, the Pratt Institute, where he gained skills in painting, and at the same time took him to Greenwich Village for classes in pottery and woodcarving.

Everett brought the same guileless approach to his artistic journeys as he did to imbibing the wilds. He was never shy about seeking teachers, heading out of the canyons now and then back to Los Angeles or San Francisco, where he'd literally knock on the doors of famous painters and photographers and ask their advice. Renowned western artist Maynard Dixon ended up giving him art lessons. Dorothea Lange invited him into her home to talk composition and lighting—even took portraits of him. Ansel Adams traded one of his own rather famous photographs for one of Ruess's block prints. The same kind of block prints Everett had taken to selling on the streets of the little towns in southern Utah he drifted through, using the money to buy food, shoes, and other provisions. Arriving on the streets of town smiling and disheveled, leading his two burros, to locals he seemed either a pauper or a prophet.

Finally, Everett liked to sing, creating what amounted to soundtracks for the different places and even weather conditions he found himself in: one soundtrack for a storm, another for hot sun, still others for "when the winds are blowing through the star-tipped pines." Today

I imagine that whatever tunes he came up with for the pines, which were so often ponderosa, there was likely often gratitude in the lyrics. On one journey, having dropped off the South Rim into the bottom of the Grand Canyon on a long and twisted walk toward Zion, he found himself in what amounted to a furnace; in one journal entry he guesses the temperature to be 130 degrees. Cowering in the shade past nightfall, he slogged back up and out of the canyon in the dark along the Kaibab Trail, the path in places so narrow and crumbling that even his burros lost their footing. Six thousand vertical feet later, he touched the North Rim. And there, he was welcomed by the cool of a sprawling old-growth ponderosa forest.

What Everett saw that day on the north rim likely wasn't much different from the forest that had so thrilled Grand Canyon geologist Clarence Dutton back in 1887. These groves weren't impenetrable, Dutton explained. In fact, he and his men could "look far beyond and see the tree trunks vanishing away like an infinite colonnade." Held within the woods was "a constant succession of parks and glades— dreamy avenues of grass and flowers winding between sylvan walls or spreading out in broad open meadows."

I think of the future wanderers of canyon country, how their own journeys will need to be threaded through the harsher conditions forged by climate change. Losing ponderosa all by itself is going to make hardscrabble foot travel—which even now teeters on the edge of danger—that much harder. There'll likely be fewer springs, as trees no longer gather up moisture from rain and snow and channel it back to the aquifers. And, of course, less shade—this in a land where shade is priceless.

When Everett was nine years old, his teacher asked everyone in the class to write a made-up story. His was about a man out in the desert, sprawled on the stinging hot sand under the twist of a Joshua tree. Everett tells how the crooked shade made fantastic patterns, "falling in sultry stripes across his weary body"—and how the man lurched as the shadows moved, trying to keep his head in bands of shade. He was an artist, the story reveals. But sadly, the artist's soul was dead within him. "He'd come to the desert to die," young Everett wrote. "Or to recover his lost ambitions."

~

The ponderosa forest where I've come to rest on this hot day, high on Thousand Lake Mountain, is actually one I've been in before. Seventeen years ago, at a time when these trees were offering me a very different kind of shelter. It was six months after my first wife, Jane, had been killed in a canoeing accident, drowning in a cold, roaring river in northern Ontario. Honoring a request she'd made a dozen years before that tragedy happened, I came here to canyon country, to Capitol Reef, to scatter her ashes. To leave some of her remains in the landscape she once described as having saved her life.

Jane suffered in her teen years with a long, nearly deadly bout of anorexia. At an especially low point in her life, a favorite cousin, knowing her love for the outdoors, suggested she check out an Outward Bound course. Established in England near the start of World War II by educator Kurt Hahn, Outward Bound was initially meant to help sailors build both mental and physical skills to help them survive stranded at sea. The results were impressive. After the war ended, a version of the program landed in the United States, tweaked to fit a favorite notion we'd had in this country since the nineteenth century: namely, that the outdoors was a powerful healing place for young people with flagging self-esteem. I've often heard Outward Bound instructors—and Jane would later become one—say that their role was mostly about helping students feel safe out in the wilds. Once that happened, they said, the wilderness did the rest.

The course she picked back then was at Capital Reef—a place she chose in part because the landscape looked nothing like her home back in the farm country of southern Indiana. She first told me about it a month after we met, during a backpacking trip in the Sawtooth Mountains of Idaho. Beyond sharing what the day-to-day experience was like, she tried to explain to me why the experience was so healing. The first breakthrough, she said, had to do with a kind of reimagining of her hardship. I remember at this point her stopping in front of me on the trail, turning around to face me, then reciting a three-line quote from Robert Pirsig's *Zen and the Art of Motorcycle Maintenance*—something shared with her and her fellow Outward Bound students on their first day in the canyons. "Physical discomfort is important only when the mood is wrong. Then you fasten on whatever thing is uncomfortable and call that the cause. But if the mood is right, then physical discomfort doesn't mean much."

She said she found over time that that this was also true for emotional discomfort. The key for her in shifting an attitude was to be totally present in the moment; and once there, to allow herself not just to face the fear or anger but to walk right into it—to feel it fully, without any need to analyze it. The key was that alignment with the present moment. And there was no place where being in the moment came so easily as out in nature. "If you can really stay with what's actually happening, then whatever irritation you've got going—maybe something was *supposed* to happen, but it didn't—loses its grip."

Jane started instructing for Outward Bound in the early 1990s, often leading middle-aged people struggling through difficult life transitions. Maybe losing a job. Or the sadness of the last kid moving out of the house. A divorce. The death of a family member. She told me in those years how out on the trail, especially in the first day or two, it wasn't unusual for some people to get restless, even seriously grumpy, wishing the rain would stop, that the day would get warmer, that the bugs would go away. But unlike when they were in bad a mood at home, it was different for them when they voiced such complaints out in nature. Out in the kind of big wilderness where they were traveling, mostly in the Rocky Mountains of Montana, the ability to control things was so obviously limited. It took little time before a person's upset about what nature was *doing* to them seemed ridiculous.

"That's freedom," she liked to say.

I was thinking a lot about that conversation when I scattered her ashes in these canyons all those years ago, then drove up here to sit with the trees. And now I'm thinking about it all over again. What would Jane have to say about the ponderosa disappearing from so much of this land, taking with them the safety and reassurance they'd provided to people for thousands of years? Is it possible to accept where we are right now, to acknowledge our mistakes, and at the same time keep walking forward, moment by moment, through our anxieties? Or will we get stuck in our anger and depression, mad or sad as hell when we think how things shouldn't have turned out like this?

Doctors and therapists who work with people with acute physical pain sometimes offer the message that things can get better not by blunting the pain or distracting themselves from it but by walking straight into it. It won't eliminate it. But sometimes pain isn't quite so quick to chase you down and pummel you if you can just stop running away.

~

Despite being in this beautiful grove of ponderosa on the top of a high mountain, when I'm in canyon country, I can't help but think of this tree growing down lower, living solo in the bottom of the canyons. The only beings who can gain a stature eight or ten stories high; nothing else even comes close. Some of these solitary trees started as seeds dropped into a crack in the rock, maybe lost from the beak of a passing bird. From that humble beginning, over a century or two or three they've risen inch by inch to soar against the slickrock. Their dusky green branches, less striking when part of an entire forest, seem more bold when the tree stands alone in such a place, a daring, audacious stroke of life in an endless warble of stone.

Ten thousand years ago, when the climate was cooler and wetter, ponderosa was fairly plentiful down lower in the canyons. But that just makes the ones who still make their homes there today, when conditions are much less friendly, all the more intriguing. Writer and

naturalist Jim Stiles considers them holy trees. He calls them *outliers*.
The kindred spirit trees of canyon country's nomads: Ed Abbey and
Everett Ruess, to be sure, but also a rucksack full of lesser-known lon-
ers, like Toots McDougald and Herb Ringer and Jack Holley.

It's for the chance to dance with one or two of those outliers that
I've decided to head west from Thousand Lake Mountain, no matter
the heat, and make for Zion National Park. Zion, much like Bryce
Canyon to the north, has ponderosa in both of its manifestations—
sprawling forests draped across the higher elevations, and then, lower
down, solitary elders. A few have managed to thrive in even more preposterous locations: like high up on the sheer face of cliffs, looking like
BASE jumpers about to leap.

My drive loosely follows that trek Everett Ruess made on foot, after
climbing out of the Grand Canyon onto the Kaibab Plateau. Zion,
too, like so much of the canyon country, was special to him. This, even
though in late summer of 1931 he was struck down in the park by a vi-
cious poison ivy infection. By the time Ranger Donal Jolley stumbled
across Everett lying along Zion's West Rim Trail on September 2, he'd
spent more than a week in serious misery. "For two days I didn't know
whether I was alive or dead," he wrote. "Ate nothing, simply writhed
& twisted in the intense heat & swarms of ants & flies. Finally, I managed to pry my lips far enough apart to insert food." Ranger Jolley got
Ruess to the hospital in St. George, where he spent five days recovering
before setting off again. Stopping first, though, to pick up his beloved
donkey, Curley, whom the Jolley family had been boarding while Everett was recuperating.

True to the premonition Everett laid out in his journal—how when
it was his time to die, he would make some last wilderness trip and not
return—he vanished in November of 1934, four years after first beginning his wild wandering in the Southwest. He was twenty years old.
Some think he met his fate in Davis Gulch, near present-day Bears
Ears National Monument—not by any misfortune of his own making
but at the hands of local rustlers who mistook him for a government

agent. Others say he drowned trying to cross the Colorado River—though given how low the water would've been that time of year, combined with his ability as a swimmer, this seems unlikely. No one knows for sure. A month before he disappeared, Everett mentioned in a letter to his parents that he was reading Homer's *Odyssey*. He'd reached the part where Odysseus is trapped in a cave by a man-eating giant and finally saves his own life by announcing himself to the giant as "Nemo," which is Latin for "nobody." After his disappearance, searchers in Davis Gulch found in two places an inscription scratched into the rock: "NEMO 1934."

Once I hit the trail in Zion, it takes about an hour of walking to find what I'm looking for: a beautiful, slightly crooked ponderosa about seven stories high tucked into a slot canyon, the frizzled green paintbrush of her canopy electric against the blue sky. Down low on one side of the tree, two surface roots wind out from the base of the trunk to hug the edges of a car-sized block of stone. An even larger root on the other side twists off in the opposite direction, finally disappearing down a long fissure in the rock, which it split open long ago in its ongoing quest for water. The rest of the ground is soft, a bed of rusty blond sand threaded here and there with morning glories and scarlet gilia. There are tracks from mule deer and pocket gophers all over the place—and a lone set, too, from a big ringtail cat. This being the only tree around, the vanilla scent oozing from its bark is faint, fading in and out against the smell of juniper and the clean, dry scent of sand.

Given that ponderosa are sun-loving trees, preferring open, sunlit spaces to grow, it's amazing trees like this one end up so robust with no more than a token couple of hours of direct sun on any given day. I ditch my daypack at the base of the tree and start poking around. In the upper reaches of the canyon, a lone turkey vulture is circling; a strong flier, to be sure, and yet the small corrections she makes to her wingtips every couple of seconds leave her looking like a kid fresh off her training wheels. When the sun clears the lip of the canyon rim and falls on me directly, the heat soars. I'm guessing it's at least a hundred degrees.

Once again it's time to give in, retreat. So I find a soft spot under the ponderosa canopy and lie down on my back in the sand.

Later in the afternoon, when shade returns, I make my way to a lovely hanging garden called Weeping Rock. When it was clear that Everett Reuss had in fact disappeared, when there were no more letters arriving from those tiny towns in southern Utah, his parents Stella and Christopher set out on a crusade to find their son. Even decades later, after all hope was gone, they still came back now and then to the towns and small villages of southern Utah and northern Arizona, hoping to at least hear of other people's encounters with him. For all those melancholy journeys, it was here, near Weeping Rock, where Stella said she most felt Everett's presence. One afternoon she found a quiet alcove, settled in beside a small trickle of water, and wrote a poem to her son. It was while sitting here in the slickrock bosom of Zion, she later said, that for the first time since she and her husband started searching for their missing son, she no longer felt the need to cry.

Wallace Stegner described Everett Ruess as "a callow romantic, an adolescent esthete, an atavistic wanderer of the wastelands, but one of the few who died—if he died—with the dream intact." Maybe more on the mark still was a line from the rabble-rousing Ed Abbey, whose words I suspect would ring true not just for Everett but for a great many people who've wandered long and deep in canyon country: "You knew the crazy lust to probe the heart of that which has no heart that we could know."

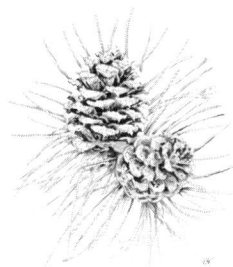

Chapter Twelve

By the time I hit the town of St. George, just outside Zion, in the driest corner of Utah, it's 107 degrees. As national park gateway towns go, this is an especially beautiful one, cradled by stunning red desert cliffs and, off to the north in the distance, the ten-thousand-foot-high Pine Valley Mountains. Not long ago, St. George was deemed the fastest-growing city in America, which may be a bit of a problem. Not only is water limited here, but weirdly, Utah has the dubious distinction of having the second highest per-capita usage in the country. In Washington County, home to both Zion and St. George, in 2020 that use was an astonishing 285 gallons per person per day, more than twice the per-person rate in nearby Las Vegas. "I do not believe Utahns have fully grasped the magnitude of what we are facing," Terry Tempest Williams writes. "We could be forced to leave. Our natural touchstones of joy will deliver us to heartbreak. Each of us will face the losses of the places that brought us to life."

The land becomes starker as I drop into Nevada—more specifically, into the heart of that vast and superbly lonely region known as the Great Basin. Sprawling across more than two hundred thousand square miles of the interior West, this is one of the few places where ponderosa often disappear from the landscape, thriving only on the high, windswept flanks of some of the more than three hundred north–south mountain ranges. Mountain ranges that Clarence Dutton—the geologist who was so enamored of the spacious ponderosa groves on the Kaibab Plateau—described as "an army of caterpillars crawling northward out of Mexico."

The growth pattern of ponderosa in the western United States forms the shape of a massive, upside-down letter U. The eastern leg is made up of trees running north out of New Mexico and Arizona for some seven hundred miles, drifting now and then onto the Great Plains in scattered sky island groves, including those in and around the Black Hills. (The Black Hills actually take their name from the ponderosa, which bear charcoal-colored trunks as young adult trees.)

Then the trees flow west, sometimes in small bands and other times in full-blown choruses, across Montana into Idaho, Oregon, and Washington. In these western reaches we find the three-needle variety of the tree, which plunges south to Mount Shasta and down the Sierras. In the middle of that big inverted U-shaped growth pattern, basically from the edge of Utah's Wasatch Mountains to the eastern toes of the Sierras, from southeast Oregon and southern Idaho south to the Mojave Desert, is the Great Basin. The biggest desert in America, and the one place in the region where ponderosa is largely absent. The arid flats here are made up of thin, alkaline soils, barely rich enough to support sagebrush and grama grasses and saltbush and Mormon tea. Still, in some of the higher places, well out of sight to most travelers, ponderosa has done just fine for itself. Indeed, high above the Great Basin in the Wah Wah Mountains of western Utah, until recently there grew what was thought to be the oldest ponderosa in the world—a massive, 940-year-old great-grandmother. She died in 2016, in the middle of a drought, meeting her end by way of an invasion of western pine beetles.

The Nevada portion of the Great Basin is sometimes cast as a place to suffer through. Nevada Highway 50, stretching more than four hundred miles across the Great Basin, has long been touted as the loneliest highway in America; westbound drivers wonder if California will ever come, while those going east keep squinting through the windshield for sight of the big blue-and-brown sign that says "Welcome to Utah. Life Elevated."

I have to say, though, that I've always found a kind of sumptuousness to this great wide open. Probably because years ago I decided to park the car, get out, and actually start walking the place—to cozy up to some of the 280 plants and animals that grow here and nowhere else in the world. But my fondness for this lovely, lonely place has become about more than just endemic plants and animals. These two hundred thousand square miles of capacious vacancy are a powerful balm for the absurd number of distractions that fill our days. Turning off the car stereo and rolling down the window in Nevada is like entering mindfulness training. As writer Ellen Meloy put it, "The Great Basin is a place where the silence has a presence, where the emptiness is full, where the vastness brings you home to yourself."

Maybe it's the profound isolation of the place—a quiet so deep you can hear the blood pulsing in your ears—but something led to the Great Basin becoming a land where strange things happen. There's the often-told tale of Nan Dixon, who vanished in 1978 while driving from her home in Grass Valley, California, to visit family in Lovelock, Nevada. Her car was found four years later, sitting in a ravine that had been thoroughly searched right after she disappeared. And then of course, there's the fanatically popular Area 51, where, among other things, the government is said to have worked to reverse engineer alien spaceships brought here after crashing to Earth.

But among the strangest stories is one that took place at the notorious Nevada Test Site northwest of Las Vegas—a place where, beginning in the 1950s, the US Army detonated nuclear devices to study the effects of the blasts. Curious about what an exploding nuclear bomb would do to a forest, yet stuck working on a test site bereft of trees, Department of Defense scientists arranged with the Forest Service to bring in 145 mature ponderosa pines from a distant canyon. About a mile from ground zero, at a place known as Frenchman Flat, excavators dug tidy rows of giant holes and into each was cemented one of the enormous trees. A twenty-seven-kiloton bomb was then exploded twenty-four hundred feet above the makeshift forest. In the grainy, colorless film of the event, you can see the trees whipsawing madly, violently buffeted back and forth in six-hundred-mile-per-hour winds. They can be seen burning. Yet incredibly, in the end, nearly all of them are still standing.

~

The Basin is playing hardball today. In truth I'm finding it hard to live in that wisdom shared by my first wife: to not get overly focused on wishing things were different. It's 116 degrees on the other side of the windshield, with a hot west wind blowing like a rocket-powered hair dryer—which makes my next stop, the forests of the Sierras, seem really appealing. So I keep pushing west.

The television show *Bonanza* may have given me my first glimpse of ponderosa as a kid. But an even deeper inkling of just how spellbinding such a forest could be came when I was in high school, lying in bed at night reading, trying to transcend the cornfields and smokestacks of northern Indiana with some help from John Muir. Even now I find it hard to think of California's forests without Muir showing up, ambling under the canopies in his wool pants and flannel shirt, a tin cup, some tea, and a loaf of French bread from Black's Hotel nested in his leather pack. And nearly always, a dog-eared copy of Emerson. In recent years, Muir—like so many early white male conservationists—has been rightly called out by many in the mainstream environmental movement for his racist attitudes. He wasn't at all kind in describing the Black people he encountered on his thousand-mile walk from Louisville, Kentucky, to the Gulf Coast; and two years after that, in California, he was calling the Miwok people of Yosemite "dismal" and "degraded."

What's a little unusual in Muir's case, though, was the change of heart he had later in life. A big piece of that change came after he spent time with the Tlingit and Stikine tribes in Alaska, coming to refer to them as "brothers," talking about how much more dignified and intelligent they were than the white settlers he encountered. Later he went so far as to talk about "the equality of all people, regardless of color or race." For that matter, Muir even had a change of heart about the Miwok near Yosemite, saying that "perhaps if I knew them better, I should like them better." Probably so. One thing he would've understood is that the desperate conditions of the Miwok when he first encountered them in 1869 was the result of the wanton rape and murder, violent relocation, and wholesale destruction of their sacred sites at the hands of local white citizens' militias.

For all his flaws, one thing Muir got right was his take on the almost aching allure of California's forests. He loved America's woodlands in general, which he surmised were "a great delight to God, for they were clearly the best he ever planted." But I suspect he would've said God's personal favorites were the same as his: those rising from the coast of California and on up into the Sierras.

By late afternoon I'm well into California, crossing Walker Pass to join the South Fork of the Kern River, reaching in another thirty miles the town of Lake Isabella. The water today, in both the river and the lake, is downright dazzling, a stark contrast to the rest of the landscape, which seems plain tuckered out. In the hills outside of town, the gray pine looks to have been in a brawl—some of it pummeled by wildfire, much of the rest by pine beetles. And if not pine beetles, then dwarf mistletoe. Not the kind of mistletoe you might find yourself kissing someone under during the holidays. This one is a tenacious parasitic plant that lives only in conifers, including ponderosa. It's awfully good at what it does, boasting traits like being able to fire its sticky seeds into the air at up to sixty miles per hour, causing them to land as much as thirty feet away. If they hit another tree, and of course in a forest they often do, they'll stick fast. In little time they begin penetrating the bark, drilling in to get all the food and water they'll ever need from their host tree. Unfortunately, mistletoe's harmful effect on conifers is amplified by hotter, drier weather. For one thing, unlike the pines themselves, mistletoe doesn't close its pores to save water during times of drought. This means that moisture from the host tree—which is how mistletoe gets supplied—runs out through the parasite and is let loose into the air, further dehydrating the tree at the worst possible time.

Meanwhile, off to the northeast of Lake Isabella, much as in Arizona and New Mexico, the pinyon trees have been ravaged by a type of engraver bark beetle known as the pinyon ips. Like most forest insects, the ips has a thing for weaker trees. Suffice it to say that in the face of climate change, life has become for them very, very good.

Of all the natural threats to these forests, it's the western pine beetle that most plagues the ponderosa. In a highly cited study from 2021, a dozen scientists from nine different research institutions estimated that climate change has led to fully a third more ponderosa trees being killed by western pine beetles in the Sierras than would've been killed by drought alone. Some climate scientists have estimated that for every 1.8-degree increase in annual temperatures, 35 to 40 percent more ponderosa die. It's not just that warmer temperatures stress the trees, making them more vulnerable to insects, though that's certainly true. It also has to do with the fact that climate change can speed up the rate at which the beetles reproduce, leading to a lot more of them being around to invade weakened trees. Beetles kill ponderosa with a one-two punch. First, they tunnel under the bark through the phloem layer, which disrupts the tree's ability to transport nutrients; second, female beetles carry a blue stain fungus that not only helps the beetles overcome the tree's natural defenses but feeds the emerging larvae. Unfortunately, it also infects the sapwood of the tree, often to a fatal degree.

Ten years ago, just under eighty thousand trees in Sequoia National Forest died from bark beetles. The following year—the third year of a drought, with the May snowpack standing at zero percent of normal—the mortality shot to more than two million. And it's more than just ponderosa in peril. It's pinyon and gray pine, along with tens of thousands of sugar pines and white fir, Jeffrey pine and incense cedars and giant sequoia. When you throw wildfire on top of thirst-related death and beetle kill, California has seen the death of more than two hundred million trees in the past decade alone. One 2019 survey of the southern Sierras suggests that almost half of the larger trees have died. No matter how old you are, whether a baby or a boomer, for the rest of your life and way beyond, some of what were once beautifully forested landscapes will instead be open, windy sprawls of grass and shrubs.

~

California, perhaps more than anywhere else in the country, is a place where our giant blunder of overeager fire suppression has met the realities of climate change in a head-on collision. Here the consequences of each are unmistakable, remaking thousands of square miles of forests, grass, shrublands, and, of course, human communities. When I was planning the California part of this journey, I intentionally avoided looking up the exact locations where the big wildfires of the past two decades wiped out vast ecological communities. I decided instead to zigzag back and forth across the Sierras somewhat randomly, picking whatever blue highways call to me, slowly making my way from Lake Isabella all the way to Mount Lassen. It's not exactly a scientific inventory of burns and beetle kills. But keeping this part of the journey loose, following my nose, seems a good way to randomly sample how the forests are faring.

Heading north out of Lake Isabella, I work my way up to Mountain Highway 99, headed for Johnsondale. If your happy place includes twisting roads—the ones that on maps look like wet spaghetti dropped on a plate—few regions will put a bigger smile on your face than the Sierras. One moment you can be wheeling along a shoulder of white granite, the next dropping down to kiss the columbine along some tumbling mountain stream. Then maybe a twisting bolt to the top of a ridgeline, there to swoon over endless reaches of rock and timber and scrub all the way to the far horizon. Some who've been in California for a long time are finding these latter views disturbing now, being not just windows into vast unbroken forests but also into miles and miles of burns and beetle kill. I'm trying to let my eyes and heart linger on both things at the same time. See if I can detect any small shards of hope resting in the in-between.

By noon the next day, I've reached the scars of the 2021 Windy Fire—a burn that roared across more than ninety-seven thousand acres, taking not only hundreds of thousands of trees but also more than a hundred structures. Every few miles, I pull over and walk out to explore what's left—and also, to look for signs of what's rising again—wandering through charred remains of ponderosa and sugar pine and giant sequoia. Today there are about a hundred giant sequoia groves left in all of California; this one fire, which in a matter of weeks combined with another burn going on in Kings Canyon and Sequoia National Parks, killed well over three thousand of the trees. Even worse, those losses came on the heels of staggering mortality the year before, thanks mostly to the Castle Fire, when between 10 to 14 percent of all remaining sequoias were lost.

By this point on the journey, some thousand miles from my starting point in north-central New Mexico, I've encountered hundreds of thousands of burned ponderosa. To say it's been sobering is an understatement—especially in those places where, even years after a burn or beetle kill, there's little sign of regeneration. Yet for me there's something uniquely harrowing about coming across burned-out sequoias. This is an astonishingly resilient tree—thirty stories high, wrapped in a protective layer of bark nearly two feet thick, able to live for three thousand years. That protective bark alone is nature at its brilliant best. Made from an ingenious weave of fiber bundles around pockets of air, the layer not only insulates the trunk from the heat of wildfires, but can compress and then rebound again if something big strikes the outside of the tree. While dating fire scars on sequoia stumps near here, dendrochronologist Tom Swetnam, along with colleagues from the University of Arizona's Laboratory of Tree-Ring Research, found a pattern of frequent healthy surface fires going back more than two thousand years. I've always had this notion of sequoia being pretty much indestructible—barring, of course, humans showing up with chain saws. But now, not so much.

In a 2024 assessment of the Windy Fire, a team of ecologists concluded that "in the absence of reforestation and other forest management activities, giant sequoia seedling densities may be insufficient to sustain giant sequoia at desired densities in the future." That clear but rather antiseptic academic speak doesn't begin to convey the awfulness of the situation. A better way to say it might be that we've screwed things up so badly that one of the heartiest trees on the planet can no longer do what two hundred million years of evolution equipped it to do. That we brought it down with tailpipes and smokestacks, with too much Smokey.

The summer of the Windy Fire, the southern Sierra experienced its driest year since 1924—this on the heels of extremely low precipitation the year before. In a forest that normally soaks up moisture from rain and snow, only about eleven inches of precipitation fell—barely over the amount needed to designate a place a desert. Did firefighters and ecologists know that 2021 was likely going to bring trouble with wildfire? They did. They knew because across the western United States, wildfire experts have long been using a wide array of tools to predict bad wildfire seasons and, through those predictions, come up with strategies for how best to respond.

Every year, spring through fall, fire managers closely track moisture levels in the forests and grasslands. They do this by collecting samples from both live and dead vegetation—from plants less than a quarter of an inch thick, like grasses, up through larger trees. A six-inch-wide log from a ponderosa pine might end up classified as "thousand-hour fuel," referring to the fact that when totally saturated by rain, it would—under the current conditions—take roughly a thousand hours for that log to dry out again and be ready to burn. Grasses, on the other hand, might go from totally soaked to ready to burn in just an hour—making them a one-hour fuel.

Each week these samples are sent in airtight containers to regional labs, where technicians dry them in special ovens. Every sample is assigned a number, determined by comparing the weight of the wood at the time it was collected to the dry weight after baking in the ovens. This number—a fundamental statistic in wildfire management—is known as the fuel moisture level. High fuel moisture levels mean any fire that starts will likely move more slowly, because the energy held in the flames has to go into drying the water in the wood before the wood will actually ignite. Low fuel moisture levels, on the other hand, predict a more rapid fire spread, especially if pushed by wind. The fuel moisture level is one of those pieces of information that helps managers decide where to station things like bulldozers, aircraft, wildland fire and brush trucks, respirators, and emergency supplies of food, water, and medicine—assembling them near those places that seem most likely to burn. That said, there are more and more years now when such fire analysis shows critical conditions across much of the West. And that makes it hard to know where to park the resources.

Beyond just fuel moisture, though, fire managers are also increasingly looking at something with the somewhat wonky name of "vapor pressure deficit." We'll just call it VPD. In simplest terms, VPD points to the difference between how much moisture is in the air at a given moment, compared to how much there *could* be if that air was totally saturated. You could have, say, a 50 percent humidity level on a 60 degree day, and you could also have that same humidity level at 90 degrees. What VPD points to is the fact that in the second example, the warmer air will take a lot more moisture to become fully saturated than would the cooler air.

When the VPD is high—in other words, when the air could hold a lot more moisture than it currently does—the atmosphere will end up drawing moisture out of plants and trees through the pores in their needles and leaves. (While ponderosa can, to an extent, open and close its pores—and really to a bigger degree than many other trees—it can't keep that up for long without shutting down the processes that keep it alive.) Naturally, the more moisture the plants lose, the drier they become. And the drier they become, the faster wildfire is likely to spread. Climate scientist Carly Phillips points out that in the forests of the American West, from 1984 through 2015 human-caused climate change was responsible for, at minimum, two-thirds of the summertime increases in VPD. During those high-VPD years, while in many places the actual number of fires didn't go up all that much, the total acres of land burned nearly doubled.

Once a wildfire gets going, the people working it will bring out still more analytical tools, each meant to help them better predict how it might behave. All kinds of numbers are crunched: the age of the trees, the density of the forest, the presence of standing dead timber from beetle kill, what every nook and cranny of the local terrain looks like, current and forecasted weather, the direction and speed of the winds. Is the humidity rising at night, as it almost always used to do, causing the fire to settle a bit, giving firefighters time to catch their breath? Or is this another time—increasingly common with climate change—when that humidity is staying dangerously low even through the dark hours?

Firefighters needed all this knowledge and more to fight the Windy Fire. It ignited on California's Tule River Indian Reservation in early September 2021, on a night that saw more than a thousand lightning strikes. After two days, the fire was about 115 acres; two days later, it was 450 acres. In a week, it was four thousand acres, having leapt across the Western Divide Highway, where I'm now walking, and into the Long Meadow sequoia grove. Evacuations were issued for the nearby communities of Camp Nelson, Sequoia Crest, and Alpine Village. In one of several heroic moves, smokejumpers scaled the magnificent Bench Tree giant sequoia, hoisting up a fire hose, which they used to put out flames in the canopy of another giant tree immediately adjacent. The fire stopped only when autumn weather rolled in. Ninety-seven thousand acres had burned, with suppression costs topping $75 million. As for the mature ponderosa groves in the footprint of the fire, mortality was about 80 percent.

I do find some evidence of regeneration, including in the Long Meadow sequoia grove, though signs of new tree life are scant where the burn was most severe. While the Park Service and other agencies are working at reforestation, again, this is an enormous task.

It's worth mentioning an odd and satisfying postscript to this particular tale of ponderosa and sequoia going up in smoke. In the seasons after the burn, when, as usual, grasses and shrubs were rising fast from the nutrient-rich ashes, it took little time for grazing mule deer to come in by the dozens to feed—which in turn, helped draw in a beautiful female gray wolf and her four pups, who went on to set up house in the fire scar. Not only did the burn site offer the grass that fed the deer that then fed the wolves, but there was no end of good choices for open den sites. Known as the Yowlunmi pack, they were descendants of a famous male wolf known as Oregon 7, who in December 2011 became the first wolf in modern history to cross into California. The little family who came to live on this burn site was thought to be the southernmost wolf pack in the state—and the first wolves in 150 years to take up life in what's now the Sequoia National Forest.

I take my usual dinner—a ridiculously fat burrito—in an unbroken reach of trees outside the burn scar, smack in the middle of a glorious mix of ponderosa, sugar pine, and giant sequoia. A Steller's jay is yakking overhead, undaunted by the heat, though he soon gets bored with me and heads off to find something more deserving of his attention. Like many of us, he and his cousins are facing what could be some big reshufflings as climate change growls on; already their populations have started to slip. The same is true for several other birds, from hairy woodpeckers to mountain chickadees to Townsend's solitaires, horned larks and ruby-crowned kinglets and yellow-rumped warblers.

Sitting against a middle-aged ponderosa—the tree eating sunlight while I eat beans—I notice again that, much as was true back in Arizona, this intense heat is changing both the feel and the smell of the forest. The perfume of sap—so much a part of the fresh pine scent in the summer woods—is weakening as the trees have less fluid to push up their trunks. In its place is the smell of old pine needles, still peppery but faint and a little stale, and this in turn wrapped in big, stuffy curtains of drifting dust. It's the dust I notice most, which leaves the forest smelling like an old, unventilated library with a stick of pine incense burning in the corner. The days of me being able to walk silently, which is fairly easy to do in deep pine-needle duff, are long gone. Now every footstep I take, or even when I just lean back on my elbow, kicks off a sharp, brickled crunch. Above me, the birds sit high up in the branches, panting; the deer, meanwhile, I imagine to be cuddled up against some tired poof of buckbrush or manzanita.

For the people who actually live among the forests of the Sierras, these woods are way more than just pleasing, more than just beautiful. They're old friends, ones that have rooted deeply into people's sense of place. Some moments the trees seem, above all else, awesome. At other times they're there to quietly pull you out of some worry or sadness, an arm around your shoulders. People who live in the Sierras don't just inhabit the forests. The forests inhabit them.

Losing trees at such a relentless rate, some never to return, is overwhelming. Some people these days are holding on by using the losses as a call to action, planting trees or collecting seeds or otherwise fighting climate change, on the chance a forest may rise again. Other people, though, are either angry or forlorn, slumping under the weight of the losses—something I, too, find myself caught up in now and then. As I am again today. In particular, I can't shake thoughts about how losing sacred trees—and so many California forests seem like holy places—was for much of human history utterly devastating.

One of the most demoralizing acts of conquest was to destroy trees deemed sacred by another culture. If eating from a forbidden apple tree was the original sin, cutting down sacred trees may have been the original blasphemy. When you besiege a city, the Old Testament book of Deuteronomy commands, you shall not destroy its trees. "Is the tree of the field a man, that it should be besieged by you?" (Ironically, during early Christianity one test some church leaders demanded of new converts—a means of proving their allegiance—was to cut down a tree sacred to their birth culture.) Restrictions against cutting down trees show up everywhere from ancient China to Greece, Europe to Egypt to Japan to much of Africa. In fact these prohibitions may well be the first laws ever fashioned to protect nature.

This hugely offensive act of messing with people's sacred trees landed in colonial America, too, by way of a dustup in New England in 1765: On August 14 a group of defiant patriots pressed around a big elm on the Boston Common, using it to hang in effigy a local tax collector working for the Crown. Before the week was out, it was being called the Liberty Tree—the gathering spot of choice for anti-British rallies. Locals festooned it with lanterns and flags, even an engraved copper plaque. It also inspired a flood of patriotic poetry and verse. This was, to put it mildly, a potent tree—which is why on September 1, 1775, as soon as the British got the chance, they set about chopping down the Liberty Tree and hacking it to pieces.

But no story of tree destruction has made me shudder more than one said to have played out nine hundred years ago, in the middle of an ancient oak forest. It involved those preeminent lovers of trees, the Celts, from a tiny woodland village near Geismar, Germany. Likely there were two or three hundred people living in the forest, much as their kin had done for centuries. The trees were enormous, especially the oaks, some of which were close to a thousand years old. The sorts of trees where it could take a dozen people standing together in a circle with their arms outstretched to reach around the trunks. The biggest of the trees grew in the heart of the forest, sporting a canopy more than ninety feet across.

The villagers had long come together under that canopy, cooling themselves on hot summer days, or taking shelter from the rain. Couples got married there, and some fifty generations of babies had been blessed under the canopy. Each year offerings were laid at the base of the trunk to honor the dead. For these people, what would become known as the "oak at Geismar" wasn't just pleasing to the eye. It was a precious relative, cherished the way a person would cherish a beloved grandmother.

Then came calamity. No one knows for sure how it went down, but it might have happened something like this: Just after sunrise on a clear morning in May, twelve strangers dressed in the robes of the Catholic church stormed up the shaded dirt path into the heart of the village. Each man carried a two-foot wooden club, which they swung freely, pummeling any villager who tried to interrupt their advance. Two of the monks, as well as the man in charge, a priest, were brandishing long knives. As the terrified villagers looked on, these strangers stormed past the grain storage bins and the potter's shed, past the spinning shop and sheep pens and on into the heart of the forest, finally stopping beside that colossal oak.

Before anyone could figure out what the men wanted, the priest tugged a tawny sheet of canvas to reveal a long-handled axe; even in the shadows, the polished blade gleamed. Sensing what was about to happen, several villagers rushed in but were quickly beaten to the ground. When the area around the tree was secured, the head priest stepped up to the mammoth oak, looked to the heavens, declared his allegiance to Almighty God, clenched his jaw, and raised the axe above his shoulder to strike a ferocious blow. Then again.

The monks traded off, round after round after round. The chopping continued well into the afternoon, past when the monks were exhausted, their robes sopped with sweat, some bending over with hands on their knees, panting. Sometime in midafternoon, the giant tree began to groan. For a few agonizing seconds it twisted sideways—and then in one loud, sickening moan, some seven thousand tons of wood crashed to the ground. The priest, wheezing, leaned the handle of the axe against his waist and scowled at the villagers. Some were weeping. But their tears only seemed to stoke his anger. He announced that there was a new, jealous God in charge—a furious God, ready to destroy anyone who refused to follow the one true faith.

The story spread far and wide. When King Charlemagne heard about the incident, he bestowed on the priest a new name: Boniface. *The doer of good.*

For months afterward, villagers passed by the carcass of that great oak, still slightly elevated by some of its limbs caught in the forks of surrounding trees. As the months went by, the limbs cracked and splintered off, one and then another, lowering the old grandmother to her final resting place on the forest floor. Two years later almost to the day, the priest returned, this time with a throng of workers. They spent months cutting the trunk and larger branches of the oak into boards, using them to erect a church in the hollow place where the great tree once stood. The Celtic villagers at Geismar tried to appear properly devout. Their lives depended on it. But the much-revered, highly educated Druids, whose name literally means "knowing the oak tree," carried on in the shadows, keeping alive the deep belief, the certainty, that the sacred spirit of the world was still living and breathing in the trees.

There's no Boniface to blame for the death of two hundred million ponderosa, some ten thousand giant sequoias, more than 350 million pinyon pines, and who knows how many sugar pines and Jeffrey pines and incense cedar. No one came for these trees as part of some demand for our submission. Yet the forces we unwittingly unleashed on the planet came at a time when we were certain that we, too, were "doers of good." Four hundred years ago Renee Descartes assured us that humanity's highest calling was to take fierce possession of nature. Today, with the costs of such control fantasies crashing in on us, that idea seems at best indecent, if not flatly obscene.

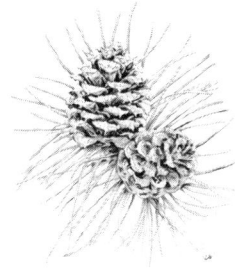

Chapter Thirteen

Though the heat continues to be overwhelming, Highway 70 is rolling north through a land fully sweetened with summer. Winding through a valley scooped out by glacial ice, now drained by the Middle Fork of the Feather River, the road leads past a world dotted with green islands of Jeffrey pine and ponderosa, in some places joined by lodgepole and sugar pine and Douglas-fir. It all plays out beautifully against the exquisite little cluster of cherry-red houses in the town of Graeagle—homes carried here in 1917 from a mill town in the Sardine Valley—a feat accomplished by literally cutting the buildings in half and then loading them on railcars, then dragging them the last few hundred yards to where they stand now. I find them a kind of exquisite box art, tiny red nests huddled under the green boughs of living pines.

I wander around Graeagle for a while, spending most of my time just staring at those green ponderosa trees. It's as true for writers as for anyone else, what psychologist Abraham Maslow said, that if the only tool you have is a hammer, everything looks like a nail. When you take a trip to chronicle something being lost, as I'm doing, something you care about, it's easy to end up seeing everything as teetering on the edge of a cliff. But beyond the unraveling that's unquestionably going on in so many places, every species in every ecosystem is right this minute going about the quiet work of trying to adjust. The urge for life that pulses on this earth doesn't disappear in hardship. The impulse of a forest to keep *foresting* doesn't ever go away. Besides doing everything they can to survive in place, right where they're growing, conifers on the whole are pushing north, or climbing upslope to cooler places in the Sierras, doing so as fast as their seeds can sprout.

The same is true for other kinds of life. Migratory birds, working to stay in sync with shifting food resources, are making spring journeys north faster than they used to, flying for longer periods with fewer stopovers. To better accommodate that new reality, in just the past forty years some species have started showing smaller body sizes and longer wingspans—changes better suited to what is now being asked of them.

Meanwhile some insects have started hiding during the heat of the day, while others—the generalists among them—have shifted to feed on plants that weren't previously part of their diets. The tricky part to understand about this shift—something made harder by our lingering Cartesian views of the world—is that this isn't about lone animals mustering independent responses. Life, which of course includes human life, has only ever thrived when individuals echo the communal reality of what ecologist Neil Evernden called "rhythms of exchange." Those exchanges are everywhere: humans providing homes to nonhuman microbes, with us giving them food while in return they break down the food we eat into usable energy. Trees are giving us oxygen while we breathe back carbon dioxide. And of course the sharing behaviors going on between the fungal networks in the soil and the roots of plants: Through photosynthesis the plant is able to provide carbon to the fungi, mostly in the form of sucrose, while the fungi breaks down essential nutrients like phosphorus and nitrogen and makes it available to the trees. We are rising and we are falling, as we always have, together.

After leaving Graeagle, I'm just past Quincy when I cross into the burn scar of the nearly million-acre Dixie Fire. This crossing will leave me with the biggest "holy shit" moment of the entire journey. Almost as far as I can see, the forest is gone, leaving the foothills and the ridgelines an open land of grass and shrubs, here and there with big sweeps of ghost trees—pines killed by fire, but yet to fall. The Dixie Fire started early in the morning on July 13, 2021. It was then that an automatic alert was triggered in the offices of Pacific Gas & Electric, indicating a power outage at Cresta Dam, west of here in the Feather River Canyon. A PG&E technician—what's known as a rover—arrived at the dam four and a half hours later. While he was able to identify two blown fuses on a nearby pole, what he didn't see on that inspection was that a sixty-five-foot-tall Douglas-fir had fallen onto the power lines. In truth it was a tree that shouldn't even have been there, given that it was only forty feet from the transmission lines and had clearly visible damage to the base of its trunk from an earlier fire. Unquestionably qualified as a so-called hazard tree. Yet it had gone undetected by line inspectors for at least five years.

When that Douglas-fir finally toppled, two of the three power lines came into contact with one another. Each line tripped its assigned fuse, cutting off the flow of energy—exactly what the system is designed to do. The third line, though, lying directly on the tree, stayed hot; and because of that, things started to burn before long. Oddly, even though the tree splayed across the line was visible from the Cresta Dam with the naked eye, most of the rest of that day nobody saw it. A company "troubleman" finally spotted it at 4:50 in the afternoon—as well as a small fire underneath, burning at a size of about twenty by thirty feet. The troubleman emptied his fire extinguisher on the burn, radioed Cal Fire, and then went back to fight the fire as best he could.

As dry as things are right now, they were even more so that July, the land frayed by relentless heat and low humidity. As far as moisture levels were concerned, it was more like October than July. To make matters worse, once the fire started to run it headed north along the west side of the Feather River through massively rugged country—places impossible to access with on-the-ground firefighters. On July 19, gusty, erratic winds began carrying embers up and over the east side of the canyon; the embers hit the plateaus on either side of the Feather River, blew into flames, and the flames started running.

About fifteen miles away, watching the sky to the northeast and trying their best to keep their stomachs out of their throats, were residents of Paradise, who three years earlier had seen their town obliterated by the Camp Fire, which left eighty-five people dead in the most lethal wildfire in California history. But the Dixie Fire was moving away from Paradise, splaying out in two very big, rapidly spreading wings. A few weeks later, those wings would sweep up and around beautiful Lake Almanor—some thirty-five miles from where that Douglas-fir toppled onto the power lines.

The Dixie Fire was not only the biggest single-source wildfire in California history but also the first to cross the Sierra divide. At its most furious, it gobbled up fifty thousand acres in a single day. More than a thousand structures were lost. Putting this one wildfire out—and it was just one of more than eight thousand fires that year—took more than six thousand firefighters, 550 fire engines, 150 aircraft, two hundred bulldozers, two hundred water trucks, and twenty million gallons of flame retardant—at a cost of $635 million. Roughly the annual budget of a decent-sized city.

As for the mature ponderosa that were lost, the one thing this superhero of the tree world sometimes has trouble doing after massive burns is reseed the next generation. Even with ponderosa's extraordinary knack for putting down deep taproots, big fires can leave the ground too harsh for new seedlings to survive. For one thing, there simply may not be enough moisture to sustain them, which means drought-resistant grasses and shrubs will end up outcompeting them. But at the same time, ponderosa seeds, which aren't as easily borne by the wind as the seeds of some trees, don't always manage to make it very far from the parent tree. Even when unburned trees are just several hundred yards away, the seeds they release may not reach the heart of the fire scar.

In just the last decade, more than 60 percent of this county, Plumas County, has been burned by wildfires, which is more than four times what burned the decade before. Much the same can be said for other nearby counties: 40 percent of Tehama County; 40 percent of Butte County; half of Trinity County; a whopping 57 percent of Napa County. Of the ten biggest fires ever to burn in California, as measured by number of acres, nine happened in the past decade.

~

To fight a wildfire, especially in the conditions created by climate change, is to wade through a constant stream of questions, many of which go far beyond the metrics of fuel moisture levels and VPD. Are people living nearby? What's the risk of embers being kicked off from the main burn, and if they are, where are they likely to land? Is the windspeed low enough to have planes drop fire retardant? Should smokejumpers go in, and if so, where? What about hotshot crews? How many ground crews are needed, and will they be safe in the places you want to send them? How many brush trucks can you get your hands on? Bulldozers? Helicopters?

Not unlike the Rodeo–Chediski Fire in Arizona years earlier, experienced firefighters watched the Dixie Fire do things most of them had never seen. It tossed out embers with such force that they started spot fires five miles away. It created gigantic heat and smoke columns that spawned massive thunderstorms, complete with lightning and fiercely erratic winds, winds that pushed fire across a hundred thousand acres in a single day.

While increasingly harsh conditions on the land are causing sleepless nights for firefighters, it's of course also an incredibly anxious situation for people living nearby. And that anxiety isn't confined just to those times when things are actually on fire. If you've been through one wildfire-related catastrophe, how do you reconcile the fact that—now more than ever—another one might be just around the corner? When I was out walking south of here, on the Trail of 100 Giants, I met a family on vacation from their home near South Lake Tahoe. When the conversation turned to wildfire, the parents recounted how, three years ago, the Caldor Fire not only destroyed their home, but turned the entire neighborhood to ashes. I confessed it was hard for me to even imagine such a loss. At that point, the mom started telling me how these days she and her friends are focusing as much as they can on the joy of watching whatever life is emerging next. Bunchgrasses and fireweed and snowbrush weaving new living carpets of green. Deer coming in to feed. Woodpeckers drumming for bugs on the burned snags. This family had found a way to take heart from the new life rising on their home ground—even though none of them, including their young daughters, would live long enough to see the return of a fully mature forest. Talking to them, it struck me that one part of resilience might consist of being able to direct your attention to a natural world that's showing you exactly what resilience looks like.

I'm thinking about that family as I drive into the town of Greenville, which was devastated in August 2021 by the Dixie Fire. To say the burn hit this handsome historic town hard is an absurd understatement. The flames quite literally roared on into town, gobbling up more than 75 percent of the community in under thirty minutes: the library, the fire station, the auto parts store, the 1860 Goss pharmacy building, the 1881 Bransford & McIntyre Store, Donnell's Music Store. Cafés, hotels, bars. Here and in the surrounding area, nearly eight hundred homes were damaged or destroyed. Surveying the damage from Main Street right after the fire, it looked like the aftermath of a carpet bombing.

As the Dixie Fire burned, one firefighter offered a brutally honest comment about the blaze: "The fire right now is just not controllable. There's really nothing we can do."

My own personal panic about seeing forests dying, and often not coming back, is mine to process. Certain other, wiser people who've gone through actual wildfire hell, like that family from South Lake Tahoe, have much to teach me. As do people like Greenville resident Ken Donnell, who in the aftermath of the Dixie Fire summed up his reality in a dozen words: "We lost one dream. Now it's time to make a new one." He and many of his neighbors are doing that—at least the ones who stayed, and to be clear, many didn't. I find a smattering of new, fire-safe homes beginning to circle what might be best described as a pop-up town, consisting of tiny bars, restaurants, even a greenhouse for gathering on cold winter days. At the edge of the community is a big fabric sign: "Greenville Lives!"

Ken Donnell's "new dream" idea could only become reality in the context of community. Community is the thing that's still working— the thing that didn't burn up in the flames. It's also the thing you can't take with you if you pick up stakes and move away. Choosing to keep growing where you're planted, when there's every good reason to cash in your chips and move away, makes even a burned-up village a sacred thing. Not that things can be exactly as they were. A lot of effort has to go into rebuilding homes so they're "fire-safe." And in some places, out past the edges of town in the most fire-prone landscapes, it might be wisest to not rebuild at all.

As I roll out of Greenville at four o'clock in the afternoon, heading for the Feather River Canyon, the temperature hits 113.

〜

Using satellite imagery from both long before and shortly after the Dixie Fire, researchers at Penn State discovered something interesting: Areas that had burned severely before, often more than once over decades, were the same areas that burned severely again. Places where the land burned with less intensity, which included lands that had been treated by prescribed burning, likewise burned that way during Dixie—this, even though wind behavior and dryness levels were unprecedented. If the trees and other plants are changed by a big fire, the new mix of life rising in its wake will recreate itself in a way that leads to similar intensities and burn patterns in wildfires to come. In other words, the land has an ecological memory.

In the fall of 2021, with the Dixie Fire still simmering, former governor Jerry Brown convened a group of fire ecologists and other scientists at his ranch in Colusa County to come up with strategies for dealing with what was clearly a new era. The result was the Venado Declaration, which called on the state to spend $5 billion a year—both to treat the existing forests with prescribed burns and thinning but also to help communities do the work needed to become far more fire-safe.

Which brings us back to something else uncovered by the Penn State researchers: Lands treated by prescribed burning burned less severely not only than unmanaged lands, but also those treated by mechanical thinning. In other words, if you prescribe burn in a thoughtful, informed way, as so many Indigenous cultures did, you'll help maintain a healthier forest, better able to recover from future fires. And this applies across almost the entire massive range of the ponderosa forest. That said, again, in reality the amount of land that can be treated through even aggressive programs like the Venado Declaration is limited. Under the best scenario, the program will likely address only about a third of the land in California that really needs it.

There's one more wild card in play here when it comes to California's wildfires—something increasingly in play all over the West—which once again has climate change written all over it. Far more than any other time in history, there are a frightening number of so-called plume-driven wildfires happening. When fires get big and hot enough, stoked by heat and wind and bone-dry fuels, they can end up creating massive convection columns, or plumes. These are essentially strong updrafts, sometimes powerful enough to pluck entire trees out of the ground and send them flying off like flaming torches. Historically most plumes topped out around six or eight thousand feet, which is plenty high enough in windy conditions to toss out burning embers—embers that can easily land in people's yards or on their roofs, where they can be sucked into venting systems, or land behind a working fire crew, starting a secondary fire that pins them between the burns.

In the Sierras, the heights of plumes have been increasing rapidly. Under the right conditions, it's no longer unheard of to see them topping out at a whopping twenty or even thirty thousand feet. That's more than enough to create some very crazy localized weather, from big winds to lightning strikes. Way spookier still, though, is that when big convection columns cool at high elevation—which happens because cool air weighs more than warm air—it sometimes brings a massive collapse. When that collapse happens, it can drive furious pulses of wind, pushing wildfire all over the place in ways that are impossible to predict. When plume-driven wildfire is in the cards, all bets are off.

~

The mood in the town of Paradise feels like a mix of friendly and sober. Perched in the western foothills of the Sierras, on the edge of a forest of Jeffrey pine and ponderosa, in 2018 this community of twenty-six thousand people was ground zero for that inconceivable tragedy known as the Camp Fire. Today the population is about ten thousand.

Whether they stayed after this tragedy or decided to go elsewhere, the images of what people saw during that event will haunt them for the rest of their lives. Frenzied nurses and doctors at the Feather River Hospital, loading fragile postsurgery patients into their cars and driving off barely ahead of the flames. People trying to escape in their cars but finding the town's two exit roads impassable, strewn with burned-out vehicles, flames seeming to come from every direction. Some could feel the searing heat through their car windows. They cursed and cried and pounded the steering wheel and frantically kept trying to phone for help. And even when they finally escaped immediate danger, they carried with them the anguish of wondering whether their moms or dads or sisters or brothers, their grandmothers or grandfathers who stayed behind until there was nowhere left to run, might well be perishing in the firestorm they could still see raging all around them.

Yet even with all that, as I saw in Greenville, a lot of the people who stayed in Paradise seem tightly bonded, like sailors who've plowed together through a terrifying storm at sea. According to social psychologist Dr. Mary M. Clare, research into how humans process collective trauma—how they ultimately navigate grief and come out the other side—shows that it has a lot to do with how well they're able to come together in the immediate aftermath of an event, helping one another tend to urgent needs. She believes that this is why, though FEMA and other agencies are essential to helping communities recover, it's important they not leave locals out of the hands-on work. The more chances neighbors have to join one another in recovery, the stronger they end up being in the months and years after the disaster.

It doesn't take me long to figure out that the people of Paradise really like this place. No, that's not quite right. They love it.

"Some never came back," a woman at Debbie's Restaurant tells me. "Those of us still here, what can I say? We're hooked on the place. Hooked together."

"Kind of a die-hard love," I suggest.

"Yep. One we can count on."

More than two thousand of the homes lost in the Camp Fire have been rebuilt, some of them nestled once again among mature ponderosa—trees that, curiously, are some of the only things that *weren't* destroyed in the inferno. The Camp Fire started in the forest when an old PG&E power line broke and sparked, not unlike the Dixie Fire. But here in town, what fueled the blaze wasn't so much trees as houses. Even homeowners who'd put serious effort into maintaining their surroundings, creating a safety zone by clearing the immediate area of trees and grass, still had their homes go up in smoke when embers landed on roofs or were sucked into attic vents.

The people of Paradise, or Greenville, are not going to be able to stop fires from coming around again. Like most Californians, they understand that if a place burned once, it's probably going to burn again. As former Cal Fire director Thom Porter put it, every acre that *can* burn in this state, will. But they do know now that there's a lot that can be done to keep the wildfires that do come around from eating structures. Fully half the homes built in more recent decades—built with techniques more suited to wildfire zones—managed to survive the Camp Fire. As for those built before 1997, only 11 percent made it through; unfortunately, as in so many communities, nine out of ten houses here were older.

Building codes were changed in big ways in the wake of the massive 2008 Humboldt Fire in Paradise, which killed two people and burned hundreds of structures, and again after the Camp Fire. Today there can be no vegetation of any kind within five feet of a building, including no overhanging tree branches. The bottom six inches of any exterior wall must be made of noncombustible material, and outbuildings have to be at least ten feet away from the home. Porches and decks less than four feet off the ground have to be enclosed, which prevents the buildup of leaves and other combustible debris.

It's worth mentioning here that in subdivisions on fire-prone land-scapes, where houses are less than about thirty feet apart, you can take every step in the book to protect your home, but if someone down the street doesn't do their part, the fire that takes their house will likely torch the one next to it, and then the one next to that will go, and eventually so will yours, in an unstoppable game of demon dominoes. In the wildfire regime we're living in now, and will be for a long time to come, when we burn, we burn together.

The new rules for building things in Paradise leave me wondering if climate change, and in particular the far more intense wildfires it's throwing our way, might be an invitation to shred some of our en-trenched fantasies about rugged individualism. These fantasies came to full bloat in the West a long time ago, showing up as a fierce resis-tance to anyone telling you what you can and can't do on your land. By bringing in new building codes, this is precisely what much of Califor-nia is doing—and has been doing for some time. This state is learning, and it's learning faster than most.

All that said, the devastating 2025 wildfires in and around Los Angeles put a spotlight on some nagging limits to the task of changing fire behavior by changing building codes. As with the town of Paradise, a great many of the homes lost in the Los Angeles area were built long before California passed that more stringent set of fire-related building codes in 2008. Indeed, in the Palisades Fire outside Los Angeles, fewer than 5 percent of the homes were built after those codes went into effect. That number was even lower in the Eaton Fire, around 3 percent—and lower still in the nearby community of Altadena, less than 1 percent. And while it is possible to do some retrofitting of old homes, that process is often expensive, far outstripping available support grants. Retrofitting a two-thousand-square-foot home, for example, can cost anywhere from $25,000 to $100,000.

That said, it's worth noting that there are increasingly other, less obvious costs showing up in this climate change–addled world. In 2024, ICF—a highly respected international technology consulting firm—estimated that a person born today in the United States will, over the course of their lifetime, incur additional costs ranging from $500,000 to $1 million from climate-change-related increases in housing, food, taxes, and reduced income. Beyond the burning of homes and the ravaging of our forests, climate change—no matter where you live—is going to be laying a lot of other deep hurt onto your children's doorsteps.

~

I'm leaving Paradise feeling overcome with images of fire. That loose, zigzag method of travel I said I wanted to employ—trying to get some kind of random sampling of how ponderosa are faring by following my nose—has led me to an astonishing collection of massive fire scars: from the Cedar Fire northwest of Lake Isabella to the Meadow Fire near Johnsondale; the Ponderosa Fire near the town of Ponderosa to the Pier Fire near Porterville. There was the KNP Complex Fire in Sequoia and Kings Canyon National Parks and the Rim Fire east of

Jamestown. And then the Caldor Fire southwest of Lake Tahoe—a burn that ate up 220,000 acres of mostly ponderosa forests, forced the evacuation of more than fifty thousand people, and destroyed over a thousand homes; it was also only the second fire to cross the Sierra Nevada Mountains, doing so shortly after the Dixie Fire, which crossed the divide twelve days before.

There was the Sugar Fire east of Portola, and the North Complex Fire south of the Feather River. The almost million-acre Dixie Fire all around and far beyond Greenville, and the Camp Fire near Paradise—again, this one being the deadliest in California history. A couple days from now, heading north, I'll pass scars from the Carr Fire, west of Shasta Lake, which caused $1.6 billion in damages and $158 million in suppression costs. All told, that's a dozen fires on this trip alone, eight of them megafires, collectively burning some two and a half million acres. If you include the fires that happened outside my route, the ones for which I didn't see fire scars, then the burned area just in California is about eight million acres. Four times the amount of land that was burned in the previous decade.

Shortly after leaving Paradise, though, moving northeast on California Highway 32 back toward the Sierras, I feel myself relaxing a bit. The forest along the road, while dry as desert toast, is nonetheless still standing tall, the trees sprawling out to wrap the near ridgelines in that delicious feathery green. They nuzzle the roadway like confidantes, close in but never crowded, a glittering blue sky shining through their upper branches. All of it suddenly seems deeply precious, reassuring at a moment when reassurance seems in short supply. I make one last stop near a pair of resplendent mature ponderosa, sisters who've stood beside one another for well over two hundred years. The cars whiz by, people coming and going about their lives, likely oblivious to me standing just off the road with my eyes closed and arms around the trees, my cheek against their bark. I say out loud what a pleasure it is to be sharing the planet with them.

When I open my eyes, a man is silently riding up the highway on his bicycle, watching me as he passes. Probably noting the sheepish look on my face, he smiles and calls out: "Always love your friends!"

"I will!" I call back. And I'm pretty sure he will too.

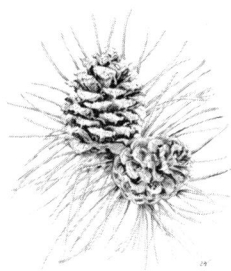

Chapter Fourteen

So many kinds of life are waning now: tens of thousands of species are at risk of extinction due to climate change alone. Incredibly, as I write this, American leaders are unraveling decades of critical sustainability successes, meanwhile slashing the very scientific research that will be critical to charting a less painful future. Under the most severe climate change scenarios, which suddenly we seem all too willing to unleash, some 30 percent of the life on Earth could go extinct before the end of the century. This centuries-old folly where we imagine we can do whatever we want to nature is about to become very costly.

It's enough to leave some people cynical, maybe lead them to conclude that ideas about fostering compassion for the Earth are hopelessly naïve. They are not. A forest employs two key strategies to survive wholesale disruption. First, it does everything it can to lessen the blunt trauma of the event. Second, it makes a priority protecting the essential ingredients it will need for recreating the community after the disaster has passed—seeds, healthy soil, pollinators—reassembling them all as quickly as possible. If it can manage those two things, the aftermath of disruption will bring growth. The current flush of arrogance in our culture is putting a great deal of life at enormous risk. But how much healing we can achieve is still very much an open question. Ultimately, it will be determined by the depth of our community.

This is one of the most essential touchstones of the forest. It is never static, any more than human society is, but rather always in motion. There are ongoing restructurings of the underground mycelium networks that transfer messages and essential substances to the trees. Rates of photosynthesis change minute to minute, and continual adjustments are being made in the pores of the leaves or needles to better match the availability of water. There are rising and falling levels of the pheromones shared between the trees, some to repel leaf-eating insects, some to synchronize reproduction with others of their kind. Individuals working together, in other words, to establish a continuity of kinship in the face of constant change.

I can feel something of this when I talk with people out on projects gathering seeds for some future forest. Or planting trees. Some whom I've spoken with explain that one way they keep going is to not get overly obsessed about the outcome of their efforts. Naturally, they'd like to think of what they're doing as an act of restoration. But the biggest part of their satisfaction seems to lie in the act of tending what they consider precious. Their reward, in other words—and at the same time, a lesson for their children—is the feeling they get from bringing honor to something they love.

The way most of us were taught to meet the world—embracing a ferocious individuality, insisting always first and foremost on rationality and efficiency—has done much to undermine our ability to live in a world that feels like home. Descartes thought that a culture moving through the world according to emotion or tradition was a dangerous "forest of error." And to be fair, in his time there were many superstitions, from misguided medical practices to the belief in the need to burn witches, that were dismantled in part with the help of rational science. But his own "forest of error" was his insistence that we had to choose sides, that we could live in only one world, the rational one or the numinous.

There's a well-worn story in addiction recovery that tells of a person walking down a street with a giant hole in the middle of it, and falling right in. Maybe they fall in the next day, too. And the next. But then, worn down and weary, there comes a time when they decide to walk around that hole. And later still, a day when they simply pick another street. The systems we've created that are feeding the ravages of climate change—and the thinking that gave rise to them—are a yawning hole in the middle of the street. We can keep falling in, blaming the Earth for having changed the game on us, or in any number of other ways casting ourselves as helpless victims. Or we can pick another street.

～

I'm back home for less than a week when I get the news that the Park Fire, just north of Paradise, has kicked up and is on the move. It will soon become the fourth largest wildfire in California history. I gasp when I see a video of hundred-foot flames roaring along both sides of California Highway 32, that sweet, tree-lined road northeast of town where I took comfort in a pair of sister ponderosa trees. The entire forest corridor goes to ash in a single day.

"Let us love trees with every breath we take, until we perish," wrote poet Munia Khan. Yes, let's. Especially now. Lately I've taken to being extra grateful for the unabashed tree lovers among us. The stockbroker on her lunch hour in Central Park, pausing at The Pond to hold her palm against an old American elm. Families out on Saturday mornings in the spring, planting saplings along a city street. Volunteer botanists and ecologists at any of hundreds of nature centers, opening people's hearts to the forest. Little kids looking up through the branches of maples, waiting for the day when they're big enough to climb. People for whom loving trees isn't just accepted. It's common sense.

Life in the Southwest will, of course, carry on. There will be red-tailed hawks and larkspur blooms and blankets of sagebrush and oakbrush; there will be brilliant light in the spring, and canyons flushed with some measure of snowmelt running to the sea. But where the ponderosa have gone away, the horizon is more lonely now. The winds seem a little more restless, no longer able, after all these thousands of years, to slow and mingle with the trees.

Today one of the most fundamental acts of repair we can make is to encourage the weaving of new stories—tales to celebrate the deep reality that this is a world of connections. Talking specifically about science, biologist Merlin Sheldrake points out that new stories always emerge from new metaphors. Four hundred years ago, in an era of prolific mechanical invention, our metaphors for nature were those of the machine. That image came to infuse everything from how we thought of our bodies to the deepest workings of society. Centuries later, during fierce industrial growth and aggressive capitalism, the metaphor became one of competition; the successful lion or bird or human or culture was by default aggressive, naturally "red in tooth and claw."

But neither our metaphors of machines nor those of all-out competition could ever capture the fact that life grows itself, tends and repairs itself, forms new relationships for the good of both individual and the collective through vast webs of exchange. "The true wood," writes John Fowles, "the true place of any kind, is the sum of all its phenomena. They are all in some sense symbiotic, being together in a togetherness of beings." This is at the heart of what makes our woodlands so uniquely precious, still an irreplaceable part of this, the only garden we've ever known.

Acknowledgments

As always, deep gratitude to my brilliant agent, Alice Martell. Also to my editor at Island Press, Rebecca Bright, for her superb ability to bring both clarity and luster to the written word. And to Mary M. Clare, for critical insights and editorial suggestions throughout the project. Special thanks as well to dendrochronologist Tom Swetnam, and to the work of forest ecologists Stephen Arno and Carl Fiedler. To the faculty of the University of Arizona and the University of New Mexico, Penn State University, the University of California Berkeley, the US Forest Service, Sequoia National Park, and Cal Fire. Also to Northern Arizona University, and Jon Ghahate at Crow Canyon Archaeological Center. And finally, to that inspiring community of citizens, ecologists, and wildland firefighters who give so much in service of this precious planet.

Selected Sources

Books

Concilio, Carmen, and Daniela Fargione, eds. *Trees in Literature and the Arts: HumaArboreal Perspectives in the Anthropocene.* Lexington Books, 2009.

Cunsolo, Ashlee, and Karen Landman, eds. *Mourning Nature: Hope at the Heart of Ecological Loss and Grief.* McGill-Queen's University Press, 2017.

Duwe, Samuel. *Tewa Worlds: An Archaeological History of Being and Becoming in the Pueblo Southwest.* University of Arizona Press, 2020.

Farmer, Jared. *Elderflora: A Modern History of Ancient Trees.* Basic Books, 2022.

Fiedler, Carl E., and Stephen F. Arno. *Ponderosa: People, Fire, and the West's Most Iconic Tree.* Mountain Press Publishing Company, 2015.

Fowles, John. *The Forest.* Ecco Press, 1979.

Harrison, Robert Pogue. *Forests: The Shadow of Civilization.* University of Chicago Press, 1992.

Haskell, David George. *The Songs of Trees: Stories from Nature's Great Connectors.* Viking, 2017.

Lawrence, D. H. *Mornings in Mexico.* Martin Secker, 1927.

Maloof, Joan. *Teaching the Trees: Lessons from the Forest.* University of Georgia Press, 2007.

Midgley, David, ed. *The Essential Mary Midgley.* Routledge, 2005.

Porteous, Alexander. *Forest Folklore, Mythology and Romance.* Read Books, 2021.

Articles

Belsky, A. Joy, and Dana M. Blumenthal. "Effects of Livestock Grazing on Stand Dynamics and Soils in Upland Forests of the Interior West." *Conservation Biology* 11, no. 2 (April 1997): 315–27.

Bloom, Khaled J., and Conrad J. Bahre. "Historical Evidence for the Upslope Retreat of Ponderosa Pine (*Pinus ponderosa*) Forest in California's Gold Country." *Yearbook of the Association of Pacific Coast Geographers* 65 (2003): 29–42.

Bonneuil, Christophe, Pierre-Louis Choquet, and Benjamin Franta. "Early Warnings and Emerging Accountability: Total's Responses to Global Warming, 1971–2021." *Global Climate Change* 71 (2021): 1–9.

Grad, Bonnie L. "Georgia O'Keeffe's Lawrencean Vision." *Archives of American Art Journal* 38, no. 3/4 (1998): 2–19.

Haffey, Collin, Thomas D. Sisk, Craig D. Allen, Andrea E. Thode, and Ellis Q.

Margolis. "Limits to Ponderosa Pine Regeneration Following Large High-Severity Forest Fires in the United States Southwest." *Fire Ecology* 14, no. 1 (2018): 143–63.

Huckell, Bruce B. "The Archaic Prehistory of the North American Southwest." *Journal of World Prehistory* 10, no. 3 (September 1996): 305–73.

Kolb, Thomas, Aalap Dixit, and Owen Burney. "Challenges and Opportunities for Maintaining Ponderosa Pine Forests in the Southwestern United States." *Forest Science* 65, no. 1 (2019): 1–14.

McDowell, Nathan G., Sanna A. Sevanto, Chonggang Xu, Park Williams, Sara Rauscher, Charles Koven, J.C. Domec, et al. "Multi-Scale Predictions of Massive Conifer Mortality Due to Chronic Temperature Rise." *Nature Climate Change* 6, no. 3 (2016): 295–300.

Merchant, Carolyn. "The Violence of Impediments: Francis Bacon and the Origins of Experimentation." *Isis* 99, no. 4 (2008): 731–60.

Mooney, Kailen A. "Tritrophic Effects of Birds and Ants on a Canopy Food Web, Tree Growth, and Phytochemistry." *Ecology* 88, no. 8 (August 2007): 2005–14.

Moore, Margaret M., W. Wallace Covington, and Peter Z. Fulé. "Reference Conditions and Ecological Restoration: A Southwestern Ponderosa Pine Perspective." *Ecological Applications* 9, no. 4 (1999): 1266–77.

Norris, Jodi R., Julio L. Betancourt, and Stephen T. Jackson. "Late Holocene Expansion of Ponderosa Pine (*Pinus ponderosa*) in the Central Rocky Mountains, USA." *Journal of Biogeography* 43, no. 4 (April 2016): 778–90.

Owen, Suzanne M., Adair M. Patterson, Catherine A. Gehring, Carolyn H. Sieg, L. Scott Baggett, and Peter Z. Fulé. "Large, High-Severity Burn Patches Limit Fungal Recovery 13 Years After Wildfire in a Ponderosa Pine Forest." *Forest Ecology and Management* 372 (2016): 91–100.

Rolston, Holmes, III. "Aesthetic Experience in Forests." *The Journal of Aesthetics and Art Criticism* 56, no. 2 (Spring 1998): 157–66.

Roos, Christopher I., Thomas W. Swetnam, and Christopher H. Guiterman. "Indigenous Land Use and Fire Resilience of Southwest USA Ponderosa Pine Forests." *Forest Ecology and Management* 430 (2018): 250–59.

Saint-Amour, Paul K. "There Is Grief of a Tree." *American Imago* 77, no. 1 (Spring 2020): 137–55.

Singleton, Megan P., Andrea E. Thode, Andrew J. Sánchez Meador, and Jose M. Iniguez. "Moisture and Vegetation Cover Limit Ponderosa Pine Regeneration in High-Severity Burn Patches in the Southwestern US." *Fire Ecology* 17, no. 14 (2021).

Swetnam, Thomas W. "Peeled Ponderosa Pine Trees: A Record of Inner Bark Utilization by Native Americans." *Journal of Ethnobiology* 4, no. 2 (1984): 177–90.

USDA Forest Service Forest Health Protection, Arizona Department of Forestry and Fire Management, and New Mexico State Forestry. "Story Map: 2021 Forest Health Conditions in Arizona and New Mexico." 2021.

Woolman, Ashley M., Jonathan D. Coop, John D. Shaw, and Jennie DeMarco. "Extent of Recent Fire-Induced Losses of Ponderosa Pine Forests of Arizona and New Mexico, USA." *Forest Ecology and Management* 481 (2021): 118724.

Yocom, Larissa, and Kristin Nesbit. "Assessing the Health and Vulnerability of Scattered Old-Growth Ponderosa Pine in Southern Utah." Final Report, Utah State University, June 18, 2024.

About the Author

Award-winning author Gary Ferguson has written for a variety of national publications, including *Vanity Fair* and *Outside*, and is the author of more than twenty books on nature and science. His memoir *The Carry Home*, which the *Los Angeles Times* called "gorgeous, with beauty on every page," was awarded "Best Nature Book of the Year" by the Sigurd Olson Environmental Institute. *Hawks Rest*—described as "dazzling" by the *San Francisco Chronicle*—was the first title to win the "Best Book" award from both the Mountains & Plains and the Pacific Northwest booksellers associations. *Decade of the Wolf*, written with Yellowstone Wolf Project director Doug Smith, was Montana Book of the Year.

Gary is a frequent keynote speaker on a variety of conservation issues, as well as a former member of the National Geographic lecture series. He and his wife, social scientist Mary M. Clare, are cofounders of the Full Ecology project, dedicated to helping people reclaim their relationship with the natural world.